S0-BVO-410

Fertilization Mechanisms in Man and Mammals

Fertilization Mechanisms in Man and Mammals

Ralph B. L. Gwatkin

Merck Institute for Therapeutic Research
Rahway, New Jersey

PLENUM PRESS · NEW YORK AND LONDON

Library of Congress Cataloging in Publication Data

Gwatkin, Ralph B L
Fertilization mechanisms in man and mammals.

Bibliography: p.
Includes index.
1. Fertility, Human. 2. Fertilization (Biology) 3. Mammals–Physiology. I. Title.
QP273.G88 599'.01'6 77-1189
ISBN 0-306-31009-0

A Division of Plenum Publishing Corporation
227 West 17th Street, New York, N.Y. 10011

Printed in the United States of America

To My Teachers

"There are three teachers:
parents, instructors, and comrades."

SEFER HASIDIM

Preface

Fertilization in mammals normally occurs within the oviduct, where it is relatively inaccessible to study. However, as a result of painstaking research, most of it carried out over the last five years, this barrier to experimentation has been largely overcome by the development of *in vitro* fertilization techniques for at least 11 different species, including man. The result has been a rapid increase in our knowledge of the physiological and biochemical mechanisms involved in the fertilization process. The aim of this book, which is an extension of my recent review of cell surface interactions in fertilization (Gwatkin, 1976), is to present a brief, but well documented, account of the new knowledge that has been attained.

Although this book deals with mammalian fertilization mechanisms, I have included some recent experiments on amphibian and invertebrate gametes to supplement the mammalian picture. This information is particularly valuable as the relatively large number of eggs available from these lower forms has advanced our knowledge of certain fertilization mechanisms beyond what is known in mammals. However, in the interest of brevity, I have omitted details of morphology and minor variations between species. For these, and other aspects not covered here, the reader is referred to the books of Austin (1965, 1968),

Austin and Short (1972), Metz and Monroy (1969), Monroy (1965), Lord Rothschild (1956a), and Zamboni (1971a).

To summarize the rapidly growing experimental literature on fertilization in a short book is extremely difficult and I hasten to apologize for any omissions, errors, and oversimplifications. My hope is that I have treated the subject with sufficient clarity to provoke the gamete biologist to new experiments, and that the general biologist and physician will find the book useful as an overall review of the subject.

Since much of my own research would have been impossible without the help of colleagues, it is my pleasure to thank John Hartmann, Garry Rasmusson, Harry Carter, Fred Andersen, Cameron Hutchison, Dorsey Williams, Harlan Patterson, and Barry Marx for their many invaluable contributions. I should also like to thank Jack Kath and Markus Meyenhofer, who prepared many of the illustrations, and Rita Pollard, who typed and retyped the manuscript.

Rahway R.B.L.G.

Contents

Chapter 1 • Early Studies 1

Chapter 2 • The Egg 5

Chapter 3 • The Sperm 13

Chapter 4 • Gamete Transport 27

Chapter 5 • Fertilization *in Vitro* 33

Chapter 6 • Sperm Capacitation 53

Chapter 7 • The Acrosome Reaction 61

Chapter 8 • Attachment and Binding of the Sperm to the Zona Pellucida 69

Chapter 9 • Penetration of the Zona Pellucida by the Sperm 81

Chapter 10 • Fusion of the Sperm with the Vitellus 87

Chapter 11 • The Prevention of Polyspermy 91

Chapter 12 • Pronucleus Formation 103

Chapter 13 • Metabolic Changes Associated with Fertilization 109

Chapter 14 • Fate of Nonfertilizing Spermatozoa and Interaction of Spermatozoa with Somatic Cells 113

Chapter 15 • Parthenogenesis 115

Epilogue 121

References 125

Index 159

Chapter 1

Early Studies

It was not until the early nineteenth century that the first penetration of a mammalian egg by a spermatozoon was observed. Prior to that time there was a great deal of misunderstanding even as to the role played by each sex in the process of reproduction. Among the ancient Greeks it was believed that the semen and the menstrual blood combined to form the embryo and fetal membranes (Austin, 1953). Aristotle suggested that the semen contributed in a nonmaterial sense, giving only form to the developing embryo, which he regarded as arising physically entirely from the female (Singer, 1950). However, opposed to this view was the Epicurean, Lucretius, who believed that the embryo was formed of the material elements of both parents. The naturalists of the Middle Ages and of the renaissance based their concepts in part on Aristotle and in part on the ideas of the Epicureans, as handed down by Galen and his Arabian followers.

By the seventeenth century, the English physician, William Harvey, 23 years after his discovery of blood circulation, published *De Generatione Animalium* (1651). On the title page is depicted the supreme Roman god, Jupiter, opening an egg from which emerge insects, fish, reptiles, birds, and man. On the egg is written "*Ex ovo omnia*" (all creatures come from an egg), a remarkable prediction, since he had not yet seen the minute eggs of mammals. Knowledge of eggs was advanced further by Malpighi (1673), a Professor at the University of Bologna, who described the development of the chicken embryo in detail. Unfortunately, he claimed to see the form of an embryo in an unseminated egg and so launched the erroneous theory of *preformation*. Mammalian spermatozoa were discovered by Leeuwenhoek six years later and it was now claimed that the spermatozoon, rather than the egg, contained the *homunculus*, or preformed organism. Thus two schools of thought arose, the *spermatists* who held that the testes of Adam must have contained all mankind and the *ovists* who held the same for the ovaries of Eve!

This state of affairs was not corrected until the eighteenth century when Caspar Wolff (1759) observed that the leaves of plants developed out of undifferentiated tissue at the growing point and argued that this could not be a process of unfolding, since the leaves were not preformed. Such development of a structure not present previously, even in rudiment, is now called *epigenesis* (origin upon). Wolff made similar observations in the developing chick, showing that the abdominal organs develop from apparently homogeneous tissue.

At this time only the large eggs of fish, amphibians, reptiles, and birds were known and it was naturally thought that the mammalian ovum must be of comparable size (Corner, 1930). Thus, when Regnier de Graaf (1672) described the growth of ovarian follicles, these were at first thought to be mammalian eggs. This error was corrected by von Baer (1827), who on examining human ovaries "discerned in each follicle a yellowish-white point . . . floating freely in the fluid." In 1843 Barry observed the spermatozoon within the rabbit egg. The male pronucleus and its union with the egg were recorded by van Beneden in 1875 and by Oscar Hertwig in 1876. Soon afterwards the significance of chromatin and the chromosomes was realized and van Beneden showed that the chromosomes of the first cleavage are derived equally from the sperm and the egg.

As histological techniques improved, including the introduction of paraffin embedding, the completion of the second maturation division and extrusion of the second polar body following sperm entry were described in the mouse by Sobotta (1895). Further morphological details of fertilization were recorded in the rat by Sobotta and Burckhard (1910), in the guinea pig by Lams (1913), in the ferret by Mainland (1930), and in the mouse by Gresson (1941). With the development of the phase-contrast microscope, observations on living eggs became possible (Austin and Smiles, 1948).

Attempts to obtain fertilization of mammalian eggs *in vitro* are recorded by many early authors, including Shenk (1878) working with rabbit gametes and Long (1912) working with mouse and rat gametes. However, failure to recognize that temperature fluctuations can lead to parthenogenesis (Pincus and Enzmann, 1936), inadequate histological criteria for fertilization, and a failure to exclude the possibility of fertilization after transfer to the oviducts of recipients make most, if not all, of these early claims doubtful (Austin, 1961; Thibault, 1969). Following the discovery of sperm *capacitation* by Austin (1951) and Chang (1951), Dauzier *et al.* (1954) incubated rabbit eggs with sperm recovered from the uterus and obtained cleavage of the resulting zygotes. In 1959 Chang transferred eggs fertilized by the procedure of Dauzer *et al.* to foster mothers and obtained the first births from *in vitro* fertilized eggs.

Chapter 2

The Egg

The mammalian egg develops from primordial germ cells characterized by their relatively large size and pronounced staining for alkaline phosphatase. These appear in the yolk sac of the developing embryo and migrate by ameboid movement to genital ridges located near the Wolffian ducts (Zamboni, 1972b; Baker, 1972). There they multiply to form oogonia, which degenerate in the medulla of the developing gonad, but they continue to multiply in the outer (cortical) regions, losing their centrioles. After several more mitotic divisions the oogonia enter meiotic prophase and differentiate into oocytes, which are characterized by prominent Golgi complexes. Cytoplasmic bridges connect the developing oocytes, and this may account for the relatively high degree of synchronization observed in mammalian oogenesis (Zamboni, 1972b). Shortly after birth the oocytes enter the diplotene stage of the first meiotic division, where they remain until the onset of puberty. In some rodents the chromosomes become so diffuse that the oocyte is spoken of as being in the dictyate stage, from the Greek word for net, *diktyon*. The diplotene and dictyate chromosomes bear faint loops, resembling the lampbrush chromosomes of lower vertebrates and invertebrates (Baker and Franchi, 1967). These are probably sites of active gene transcription.

Processes emanating from the follicle cells surrounding the oocyte form desmosomal contacts with its surface (Stegner and Wartenberg, 1961). These desmosomes probably transmit ions and nutrients to the developing oocyte (Andersen, 1974). It has been claimed that even such high molecular weight substances as bovine serum albumin may be transferred into ovarian oocytes (Glass, 1970). The follicle cell processes persist until just before ovulation (Figure 1).

Material which stains for both protein and polysaccharide develops within the cytoplasm of both the oocyte and the surrounding follicle cells. This material appears to be secreted into the space between them and eventually forms the translucent egg envelope, the zona pellucida (Braden, 1952; Odor, 1960; Stegner and Wartenberg, 1961; Jacoby, 1962; Seshachar and Bagga, 1963; Hope, 1965; Norrevang, 1968). The thickness of the zona pellucida varies between 3 and 15 μm in different species of mammals (Austin, 1961). In the ovulated mouse egg the zona pellucida is 5 μm thick and constitutes 19% of its dry mass (Loewenstein and Cohen, 1964). The fragmentary evidence available indicates that the zona pellucida is composed of proteins and carbohydrates (Braden, 1952; Stegner and Wartenberg, 1961; Jacoby, 1962; Loewenstein and Cohen, 1964; Austin, 1968; Pikó, 1969) possibly in the form of glycopeptide units stabilized by disulfide bonds (Gould *et al.,* 1971; Inoué and Wolf, 1974 and Oikawa *et al.,* 1974) and possibly hydrophobic interactions or salt linkages (Nicolson *et al.*, 1975). In view of the significant contribution of glycopeptides to the composition of the zona pellucida, it is not surprising that this structure is strongly antigenic (Glass and Hanson, 1974; Shivers and Dudkiewicz, 1974) and is able to bind a variety of plant lectins (Oikawa *et al.*, 1973, 1974; Nicolson *et al.*, 1975).

The zona of ovulated eggs is freely permeable to large molecules such as ferritin and horseradish peroxidase (Hastings *et al.*, 1972), to immunoglobulin M (Sellens and Jenkinson, 1975), and to viruses (Gwatkin, 1967). The earlier report of Austin and Lovelock (1958) that the zona pellucida of rat and rabbit eggs is

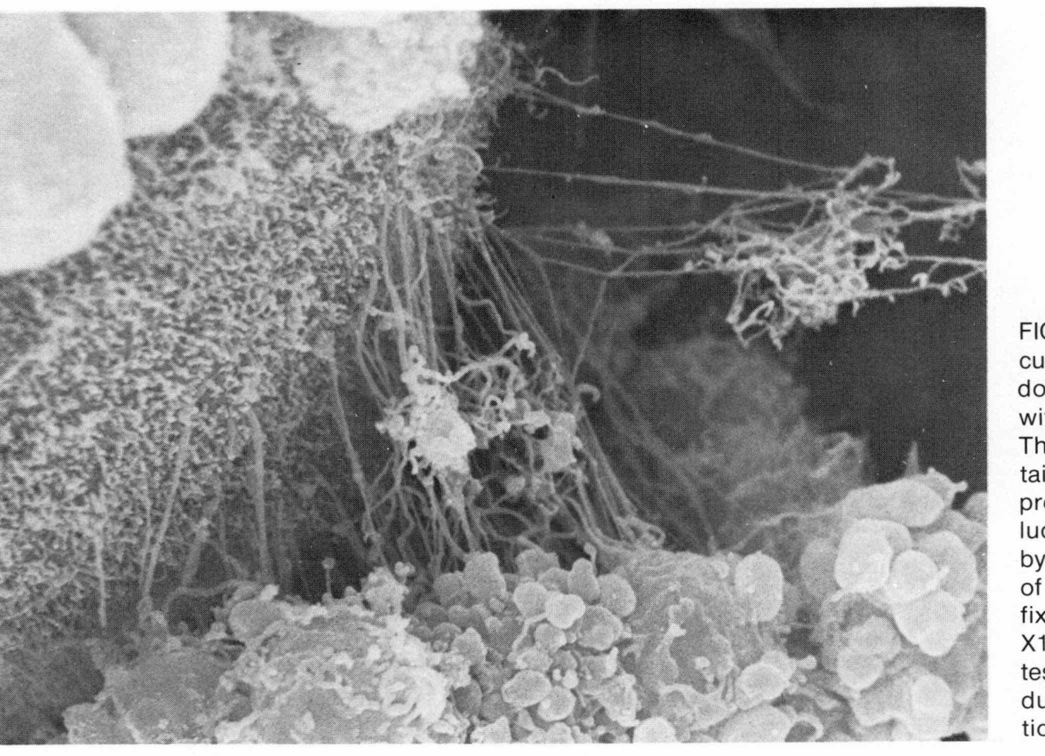

FIGURE 1. Processes from cumulus cells (top) pass downward to make contact with the surface of the egg. This preparation was obtained from the follicle of a proestrus rat. The zona pellucida has been dissolved by incubation in a solution of ATP and trypsin prior to fixation in glutaraldehyde. X1200. Photograph courtesy of P. F. Kraicer (reduced 10% for reproduction).

not permeable to heparin should probably be reinvestigated. Formation of the zona pellucida has been observed in organ cultures of fetal mouse ovaries (Blandau and Odor, 1972). The injection of 8-week-old mice with *N*(acetyl-^{3}H)-D-glucosamine labels the zonae pellucidae of their oocytes (Oakberg and Tyrrell, 1975).

The follicle cells eventually begin to secrete fluid, the antrum forms, and the egg becomes displaced to one side of the enlarging follicle. Cytoplasmic projections from the follicle cells remain to traverse the zona pellucida, and microvilli form on the surface of the oocyte. Blebs, produced by the bilaminar envelope of the oocyte nucleus develop into vesicles which later fuse to become parallel arrays of membranes (Andersen, 1974). These probably represent a specialized part of the endoplasmic reticulum. The oocyte mitochondria exhibit peripherally displaced cristae (Szollosi, 1972). This situation does not change until the morula stage, when the mitochondria elongate and develop the shelflike cristae typical of the mature organelles.

The vesicles and tubules of the Golgi complexes fill with dense material and coalesce to form large vacuoles (300–500 nm in diameter) which migrate to the oocyte periphery (Szollosi, 1967). These are the cortical granules, rich in glycoprotein (Fléchon, 1970), which play an essential role in the block to polyspermy. In the mouse, migration of the cortical granules to the periphery of the egg has been observed to continue after ovulation (Zamboni, 1970). In the recently ovulated hamster egg there are 8000 to 15,000 peripheral granules, 0.5 to 1.0 per μm^2 (Austin, 1956a). During oogenesis the number of oocytes declines drastically (Baker, 1972). Such *atresia* may occur at any stage of follicular development. Its cause is unknown.

Emergence of the oocyte from arrest in the diplotene or dictyate stage of prophase occurs just prior to ovulation (Franchi *et al.*, 1962), and this is probably in response to a rise in circulating luteinizing hormone (LH) (Ayalon *et al.*, 1972). The cause of the arrest and how it is overcome are not fully understood (Baker, 1972). Isolated oocytes cultured *in vitro* mature spontaneously to second metaphase (Pincus and Enzmann, 1936; Edwards, 1962; Gwatkin and Haidri, 1973; Bae and Foote, 1975), but they fail to do so when cultured within intact Graffian follicles unless they are stimulated by the addition of gonadotropins to the medium (Baker and Neal, 1970; Tsafriri *et al.*, 1972; Gwatkin and Andersen, 1976b). These experiments suggest that oocyte maturation is controlled at the follicular level. Other studies on hamster (Gwatkin and Andersen, 1976b) and pig (Tsafriri and Channing, 1975; Tsafriri *et al.*, 1975, 1976) oocytes indicate that the cumulus oophorus and the granulosa cells secrete a peptide of molecular weight 1000 to 10,000, which may be responsible for the arrest of the oocyte at the germinal vesicle stage. Reversal of the effect of the follicular fluid inhibitor and stimulation of the rate of maturation of the uninhibited oocyte by LH suggest that this protein hormone acts on the oocyte to enable it to overcome the inhibitory effect. A diminution in the concentration of the inhibitor as ovulation approaches may also contribute to the overcoming of oocyte arrest (Gwatkin and Andersen, 1976).

Protein hormones are known to alter the permeability of cell membranes (Csáky, 1973) and the permeability of the mouse oocyte to [^{14}C]-leucine is known to change during maturation (Cross and Brinster, 1974). Hamster ovarian oocytes begin to bind fluorescein-labeled wheat germ agglutinin after maturation, indicating a change in the distribution of cell membrane glycoproteins (Yanagimachi and Nicolson, 1974). Such changes in the oolemma would be expected to alter the response of the egg to the oocyte maturation inhibitor. In lower forms a similar situation pertains. The oocytes of the surf clam (*Spisula solidissima*) are induced to mature by exposure to the divalent ionophore, A23187 (Schuetz, 1975), which is known to alter the structure and transport properties of cell membranes (Schaffer *et al.*, 1974). In amphibian oocytes the application of progesterone externally, but not by injection, will induce maturation (Smith and Ecker, 1971). This observation implies an action of the hormone on the oocyte membrane. Such membrane effects may in turn exert an effect on the nucleus via a cytoplasmic mediator. Such a substance has not been demonstrated in mammalian oocytes, but Masui and Markert (1971) have shown that the cytoplasm of hormonally stimulated amphibian oocytes injected into untreated oocytes will induce their maturation. The active factor appears to be a Mg-dependent, Ca-sensitive protein (Wasserman and Masui, 1976).

After ovulation, meiosis of the oocyte is again arrested at Metaphase II. Figure 2 shows the structures of the ovulated mammalian egg that are known to be involved in the fertilization process. In most mammals the recently ovulated egg is surrounded by several thousand cumulus cells (Figure 3), relatively widely separated from one another and embedded in a gelatinous matrix (Zamboni, 1970). These cells are capable of secreting estrogens and progesterone (Nicosia and Mikhail, 1975), and when allowed to attach to coverslips they display prominent blebs and microvilli (Gwatkin and Carter, 1975). Information

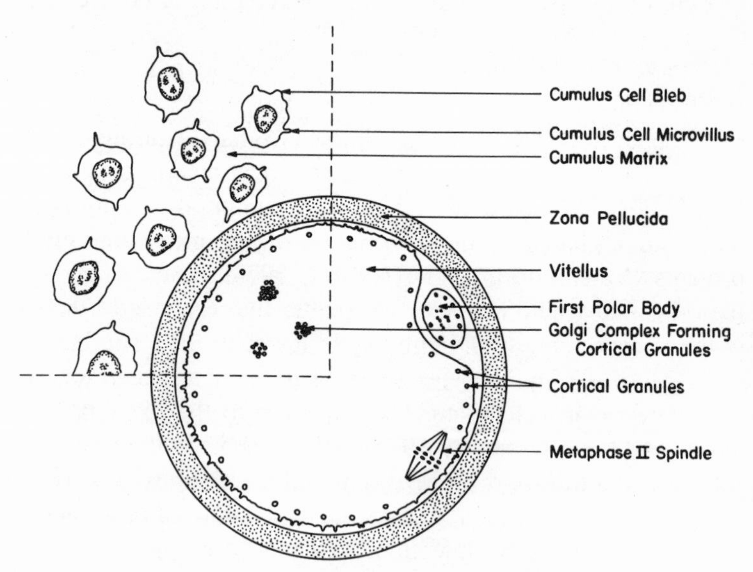

FIGURE 2. Cross-sectional diagram of a mammalian egg with its associated cumulus oophorus (Gwatkin, 1976).

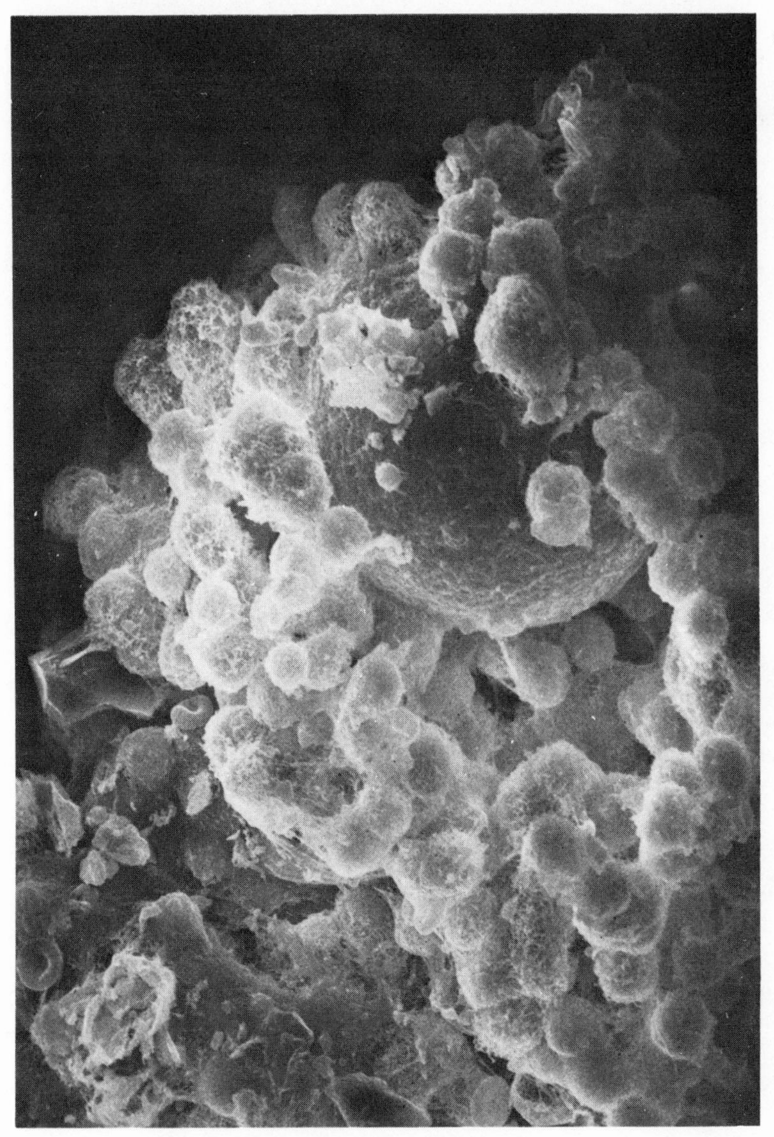

FIGURE 3. Scanning electron micrograph of a golden hamster ovum surrounded by cumulus cells. Some of these became detached during processing to reveal the sponge-like zona pellucida of the egg. X400 (reduced 10% for reproduction). (Gwatkin and Carter, 1975.)

on the chemical compositions of the cumulus matrix is limited, but hyaluronic acid is considered to be the major component (Pikó, 1969).

The ovulated egg is susceptible to aging in the oviduct (Szollosi, 1975). This is manifested either by a reduced ability of the cortical granules to fuse with the oolemma or their abnormal extrusion in blebs of cytoplasm. These abnormalities may reflect changes in the egg membrane, which include swelling and irregularity. The meiotic spindle, which is located paratangentially in recently ovulated eggs, may rotate and move to a central position. Chromosomes are easily lost from the spindle and may form micronuclei. This may be because centrioles are not present to anchor the spindle, their place being taken by loosely associated centriolar settelites which can readily separate from one another.

The investments surrounding the egg constitute major mechanical barriers which the sperm must traverse in order to make contact with the egg. Thus, spermatozoa must first interact with the cumulus cells (Chapter 6), then attach and bind to the exterior of the zona pellucida (Chapter 8), before penetrating the zona to interact directly with the egg plasma membrane. This is then followed by fusion between the plasma membrane of the penetrating spermatozoon and the egg plasma membrane. Once fusion has occurred, penetration of additional spermatozoa is blocked by discharge of the cortical granule contents into the perivitelline space (Chapter 11). These act on the zona pellucida, and perhaps also on the egg plasma membrane, to block further sperm binding and penetration. Eventually the egg plasma membrane is partly reconstituted from the membranes of the cortical granules which have fused with it, and this reformation is probably also responsible for the block to the penetration of additional spermatozoa into the vitellus.

Chapter 3

The Sperm

The male gametes also develop in the embryo from primordial germ cells, but unlike the oocytes they multiply in the inner region (medulla) of the developing gonad, forming spermatogonia (Gier and Marion, 1970). With the onset of puberty these cells divide to form primary spermatocytes which then undergo meiosis to form haploid spermatids (Courot *et al.*, 1970). The spermatozoa differentiate from these spermatids by a complex process termed *spermiogenesis* (Clermont, 1967). During this process the lysine-rich somatic histones are replaced rapidly by those rich in arginine and cystine. Feulgen staining decreases (Gledhill *et al.*, 1966), and there is an increase in the stabilization of DNA against heat denaturation (Ringertz *et al.*, 1970). The formation of disulfide bonds also contributes to the stability of the deoxynucleoprotein (Bedford and Calvin, 1974). The nucleus condenses, in part due to the removal of water. This may occur through the increased number of pores that appear at this time (Fawcett, 1975). These pores may also provide for histone exchange.

During spermiogenesis small PAS-positive proacrosomal granules appear within the Golgi apparatus, fuse, and extend around the nucleus to form the acrosome (Susi and Clermont, 1970). Concurrently, the tail develops from the centrioles, the mitochondria migrate around the midpiece, and the cytoplasm is cast off, remaining in the immature sperm as a small droplet surrounding the midpiece (Clermont, 1967). Discarding of the cytoplasm and associated ribosomes means that spermatozoa are incapable of protein synthesis or repair. The cytoplasmic droplet contains a number of hydrolytic enzymes, including nucleases (Dott and Dingle, 1968; Garbers *et al.*, 1970), and may represent a means by which these potentially harmful enzymes are excluded from participation in fertilization. The acrosome may thus be regarded as a specialized lysosome that has evolved for penetration of the egg investments (Hartree, 1975).

The sperm are eventually released from the Sertoli cells, apparently in response to LH (Burgos *et al.*, 1973), and they then pass into the epididymis. There the cytoplasmic droplet is discarded. The nucleus condenses further and the net negative charge on the sperm surface rises (Yanagimachi *et al.*, 1972). Some sperm deteriorate and are destroyed by macrophages (Roussel *et. al.*, 1967). The remainder acquire the ability to fertilize the egg as they pass through the corpus to the cauda epididymis (Orgebin-Crist, 1969). In this respect mammalian sperm differ from those of many invertebrates and certain nonmammalian vertebrates which are functionally competent when released from the Sertoli cells in the germinal epithelium of the testis.

The mature human Y-spermatozoon can be distinguished from the X-spermatozoon by a fluorescent spot, the so-called F-body, which appears on Y-sperm when smears are treated with 0.5 to 1.0% quinacrine for 20 min (Barlow and Vosa, 1970; Sumner *et. al.*, 1971). Sexing of the sperm of other mammalian species, which do not display an F-body, is not possible at present.

The mature sperm is about 50% water (compared to 80–90% for most other mammalian cells). Its length varies from about 40 to 250 μm. The dry weight of a bull sperm is about 16.5×10^{12} g (Bhargava *et. al.*, 1959). Its volume is approximately 30 μm^3 or about 1/20,000 of the volume of the bovine egg (Bishop and Walton, 1960).

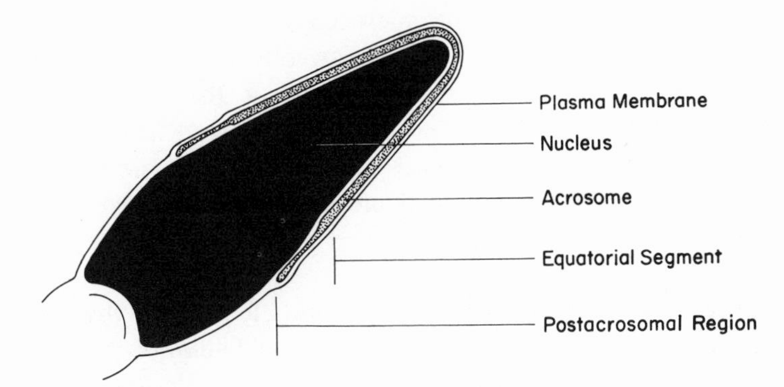

FIGURE 4. Cross-sectional diagram of a human sperm head (drawn from Bedford *et al.*, 1973).

Several recent publications have described the ultrastructure of mature mammalian spermatozoa (Bedford, 1967; Fawcett, 1970; Pedersen, 1969; Pikó, 1969; Yanagimachi and Noda, 1970d; Zamboni *et al.*, 1971). The head structures which are important for fertilization are shown in Figure 4. The plasma membrane over the acrosome is covered with tubules and vesicles (Phillips, 1975). Sperm agglutination by plant lectins indicates the presence of terminal saccharides on the sperm plasma membrane (Edelmann and Millette, 1971; Nicholson and Yanagimachi, 1972). Concanavalin A has been observed to bind to the plasma membrane of mouse sperm primarily over the acrosome, indicating the presence of a particularly large number of terminal *N*-acetyl-D-glucosamine residues in this region (Edelman and Millette, 1975). Failure of rabbit and hamster sperm to agglutinate with the lectin of *Ulex europeus*, which specifically binds L-fucose, suggests that, if present, this particular sugar is not in a terminal position (Nicolson and Yanagimachi, 1972). Characteristic but different binding patterns have been identified for other lectins and for virus particles (Ericsson *et al.*, 1971; Nicolson and Yanagimachi, 1972; Gall *et al.*, 1974). The existence of region-specific plasma membrane antigens has also been demonstrated (Henle *et al.*, 1938; Hjort and Brogaard, 1971; Johnson and Edidin, 1972). Regional specialization in the organization of the sperm plasma membrane has also been shown by electron microscopy following freeze-fracture (Friend and Fawcett, 1974; Fawcett, 1975; Koehler and Gaddum-Rosse, 1975). Over the acrosome the plasma membrane exhibits a highly ordered arrangement of particles in crystalline arrays. In the midpiece the particles are arranged in strands, while in the principal piece the arrangement is relatively random.

Labeling of rabbit epididymal sperm plasma membranes with ferritin-conjugated *Ricinus communis* lectin (specific for D-galactose residues) at 0°C, followed by incubation of the sperm at 37°C, results in a clustering of the label at the postacrosomal region but not elsewhere on the sperm surface (Nicolson and Yanagimachi, 1974). Such clustering can be understood from the Fluid Mosaic Model of the cell membrane. In this model the membrane is regarded as a mosaic of integral proteins noncovalently linked to a bilayer of lipid, including phospholipid (Singer and Nicolson, 1972; Singer, 1974). Some of these integral proteins are embedded only in the outer lipid layer, while others extend to the interior of the cell (Fig. 5). At least some of the integral proteins possess terminal oligosaccharides on their exterior poles (Nicolson, 1973). The lectin is thought to cross-link these oligosaccharides, pulling the proteins into groups. Thus, clustering may be regarded as a measure of the ease of mobility of the proteins in the lipid. Limits on such mobility are thought to be imposed by the fluidity of the membrane lipids and by the attachment of the integral proteins to peripheral proteins located both on the outside and on the inside of the membrane (Nicolson, 1973). The inner surface peripheral proteins may also be linked to microfilaments and microtubules, which further limit and control the movement of the integral proteins (Berlin and Ukena, 1972; Wunderlich *et al.*, 1973). Lectin clustering in the postacrosomal region of the plasma membrane of the sperm suggests that movement of the integral proteins in this region may be subject to less restriction than in other regions of the sperm head. This phenomenon may be significant, since fusion of the sperm with the vitellus occurs at the

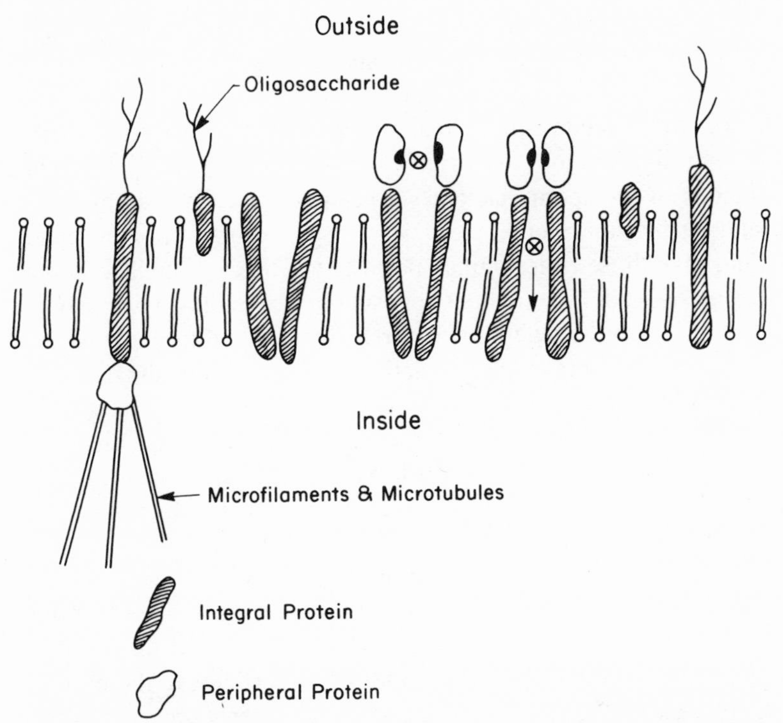

FIGURE 5. Cross-sectional diagram of the Fluid Mosaic Model of the cell membrane, showing mobile proteins (some bearing terminal saccharides) in a bilayer of lipid. Movement of the integral proteins is restricted by peripheral proteins linked to microfilaments and microtubules. The membrane proteins may form pores for the transport of hydrophilic molecules across the lipid bilayer (after Nicolson, 1973, and Singer, 1974).

postacrosomal region (see Chapter 10 and Yanagimachi *et al.*, 1973). While nothing is known concerning the linkages which may exist between proteins in the sperm plasma membrane, significant changes in the lipid composition of spermatozoa have been reported during their movement through the epididymis (Dawson and Scott, 1964; Crogan *et al.*, 1966; Quinn and White, 1967; Scott *et al.*, 1967; White, 1973; Terner *et al.*,

1975). In the ram, bull, and boar, membranes of testicular spermatozoa have a higher lecithin and cholesterol content and a higher ratio of saturated to unsaturated fatty acids than spermatozoa from cauda epididymal or ejaculated spermatozoa. Although localization of different classes of lipid to different regions of the sperm has not yet been attempted, these findings suggest that the membrane lipids in spermatozoa may be more fluid than those of testicular spermatozoa. Recent studies on the fusion of model membranes (phospholipid vesicles) have shown that an essential requirement for fusion is fluidity and that increases in membrane lecithin and cholesterol reduce the capacity for fusion (Papahadjopoulos *et. al.*, 1974). If similar conditions apply to fusion involving the sperm plasma membrane, then the above changes in lipid composition accompanying sperm maturation in the epididymis may exert an important influence on the fusion capacity of the sperm.

Changes in the distribution of negative charges on the sperm surface during epididymal maturation have been demonstrated by Bedford *et al.*, (1973) who showed that the binding pattern of colloidal iron hydroxide (CIH) particles to the plasma membrane of rabbit, monkey, and human sperm changes as the sperm pass from caput to cauda. These studies have also shown differences in affinity of various regions of the sperm surface, e.g., more CIH particles bind to the postacrosomal region than to the acrosomal region of monkey sperm. Regional differences in CIH binding have also been detected in hamster spermatozoa (Yanagimachi *et al.*, 1972, 1973). Changes in the nature and distribution of surface changes on the plasma membrane of the maturing sperm detected at the ultrastructural level are probably correlated with differences in the electrokinetic mobility in electric fields displayed by sperm populations taken from various regions of the epididymis (Bedford, 1963). These changes are not simply due to aging of the sperm, since rabbit spermatozoa retained in the caput epididymis by ligating the corpus epididymis fail to develop the charge pattern characteristic of caudal spermatozoa (Bedford *et al.*, 1973). Whether these changes result from absorption or insertion of secreted epididymal products into the sperm plasma membrane or from the unmasking or chemical alteration of existing plasma membrane components remains to be determined.

Material within the acrosome exhibits fine striations with a regular periodicity (Figure 6), which may represent paracrystalline arrays of enzymes (Phillips, 1972). Hyaluronidase, the enzyme believed to facilitate transport of the spermatozoon through the cumulus matrix (Srivastava *et al.*, 1965), has been located primarily in the peripheral and anterior portions of the acrosome of several mammalian species, using fluorescein and peroxidase-labeled antibodies (Fléchon and Dubois, 1975; Morton, 1975; Gould and Bernstein, 1975). About 50% of the hyaluronidase is released when the plasma and outer acrosomal membranes are removed by freezing and thawing ram sperm (Brown, 1975). A molecular weight of 62,000 has been reported for ram sperm hyaluronidase (Yang and Srivastava, 1974b). Mammalian sperm hyaluronidases appear to be immunologically tissue and species specific (Metz, 1972, 1973).

In addition to hyaluronidase, the acrosome contains several proteases (Allison and Hartree, 1970; Koren and Milkovic, 1973). The best characterized is the trypsin-like protease, *acrosin*, the enzyme which is thought to dissolve a channel in the zona pellucida to permit access of the spermatozoon to the vitellus. Fritz *et al.* (1975b) have estimated that a boar sperm contains 6×10^5 acrosin molecules. Acrosin has been located histochemically in the acrosomes of rabbit sperm using fluorescein-labeled trypsin inhibitors (Stambaugh and Buckley, 1972) and silver proteinate (Stambaugh *et al.*, 1975). A more specific localization has been achieved in bull sperm using fluorescein-labeled antiacrosin antibody (Garner *et al.*, 1975). Using horseradish peroxidase-labeled antibody, Morton (1975) showed that ram sperm acrosin is concentrated primarily in the equatorial and inner regions of the acrosome.

When washed ram sperm are suspended in 0.25 M sucrose, freeze-thawed, diluted, and agitated in a vortex mixer, the outer

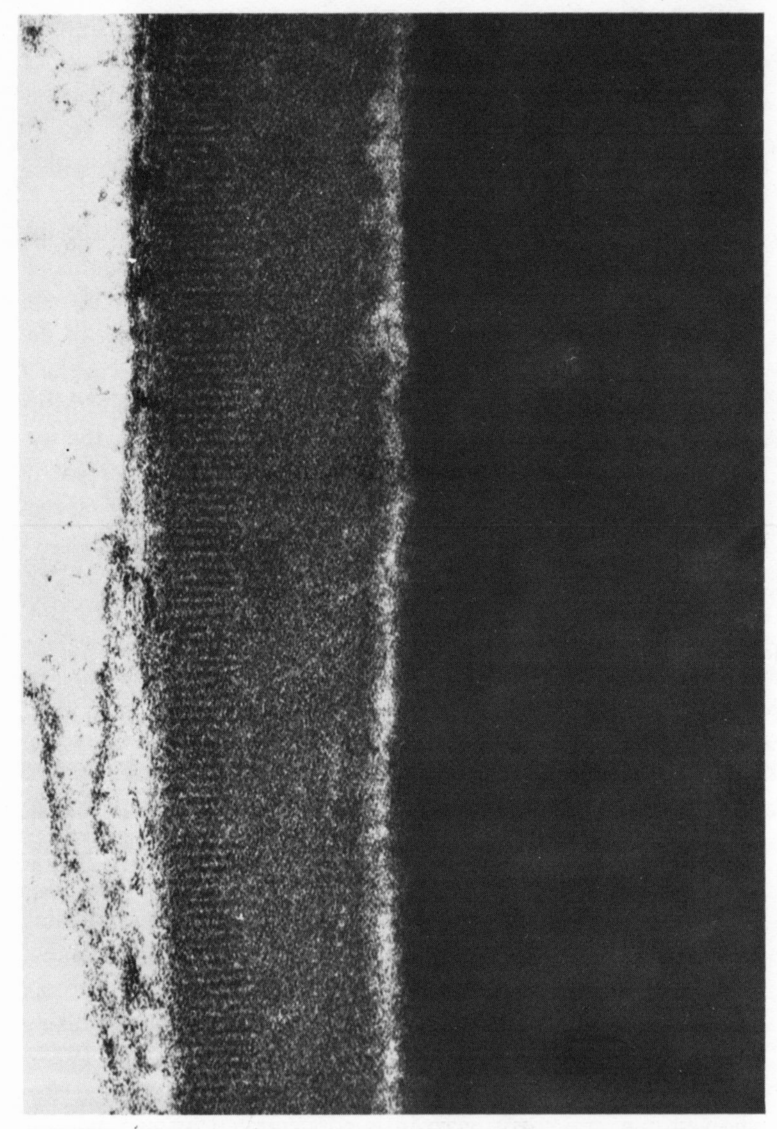

FIGURE 6. Transmission electron micrograph of a section through a portion of the rat sperm acrosome. The acrosome contents exhibit a 4.2 nm periodicity. X120,000 (reduced 10% for reproduction). (Phillips, 1972.)

acrosomal and overlying plasma membranes are removed (Brown and Hartree, 1974). Suspensions of such denuded sperm exhibit proteolytic (presumably acrosin) activity at pH 8.2 on a synthetic protease substrate, *N*-α-benzoyl-L-arginine ethyl ester, almost all of the activity being sperm bound probably as a peripheral protein on the inner acrosome membrane (Brown *et al.*, 1975). When Ca^{2+} ions are added, the bound acrosin undergoes a conformational change and is solubilized (Brown and Hartree, 1976). Since the fertilizing sperm plows a narrow furrow in the zona pellucida and soluble acrosin destroys the receptor-for-sperm in the zona pellucida (Gwatkin, unpublished observations), Brown and Hartree (1976) have concluded that acrosin is not released during fertilization but remains bound to the inner acrosomal membrane. Other reports on acrosin have failed to consider that the solubilized enzyme may be an artifact of isolation. This is an important point, since the properties of the soluble enzyme differ from those of the bound form, e.g., inhibitors isolated from ram acrosomes and ram seminal plasma inhibit soluble acrosin but have a negligible effect on bound acrosin (Brown and Hartree, 1976).

Meizel and Mukerji (1975) obtained solubilized acrosin from rabbit epididymal sperm by acid extraction and observed that at alkaline pH it undergoes a gradual autocatalytic activation which is regulated by cations. They concluded that acrosin exists as a zymogen, *proacrosin*, and requires activation before it can function in fertilization. Rabbit proacrosin was shown to have an apparent molecular weight of 73,000 by column chromatography. Early in activation this molecular weight remains unchanged, indicating that the activation process requires only a minor alteration of the molecule. Later, 38,000-molecular-weight acrosin is formed, suggesting that proacrosin may be a dimer. Polakoski (1974) has provided preliminary evidence for the presence of proacrosin in extracts of ejaculated boar sperm. Recently, Schleuning *et al.* (1976), who had previously isolated boar acrosin from stored sperm, reported that if they used freshly collected semen they could isolate an inactive form of the enzyme. Incubation under mildly alkaline conditions and in the presence of Ca^{2+}-induced autoactivation. Inhibition of this process by *p*-aminobenzamidine suggested that the activation was brought about by acrosin itself or by some other protease in the preparation.

Limited proteolytic cleavage of proacrosin could account for the discrepancies which have been reported in the molecular weight of acrosin. The molecular weight of rabbit acrosin has been reported as 59,000 (Stambaugh and Buckley, 1969) and 22,000 (Stambaugh and Smith, 1974). A molecular weight of 30,000 (Polakoski *et al.*, 1973) and 38,000 (Fritz *et al.*, 1975b) has been reported for boar acrosin. Human soluble acrosin has been given molecular weights of 30,000 (Zaneveld *et al.*, 1972) and 76,000 (Gilboa *et al.*, 1973).

Proteolytic activity, presumably soluble acrosin, can be demonstrated by smearing sperm suspensions on gelatin films and then incubating them in a moist atmosphere (Gaddum-Rosse and Blandau, 1972; Penn *et al.*, 1972). After a few minutes, halos of gelatin lysis appear around the sperm heads and gradually increase in size (Fig. 7). Development of these zones is inhibited by incorporating trypsin inhibitors in the gelatin (Wendt *et al.*, 1975).

Soluble acrosin closely resembles pancreatic trypsin in amino acid composition (Stambaugh and Smith, 1974), activity on synthetic substrates (Zaneveld *et al.*, 1972), response to inhibitors (Fritz *et al.*, 1975a), and in a requirement for Ca^{2+} (Polakoski and McRorie, 1973). However, acrosin does not appear to be identical to pancreatic trypsin since the trypsin-chymotrypsin inhibitor, HUSI-I from human seminal plasma, fails to inhibit soluble human or boar acrosin (Fritz *et al.*, 1975a).

The secretions of the male and female genital tract have been shown to contain several acid-stable inhibitors of trypsin and soluble acrosin (Fritz *et al.*, 1975a). These are of low molecular weight (6000–12,000). An example is HUSI-II from human seminal plasma, which inhibits soluble human and boar acrosin. Acrosin inhibitors have also been demonstrated in ovi-

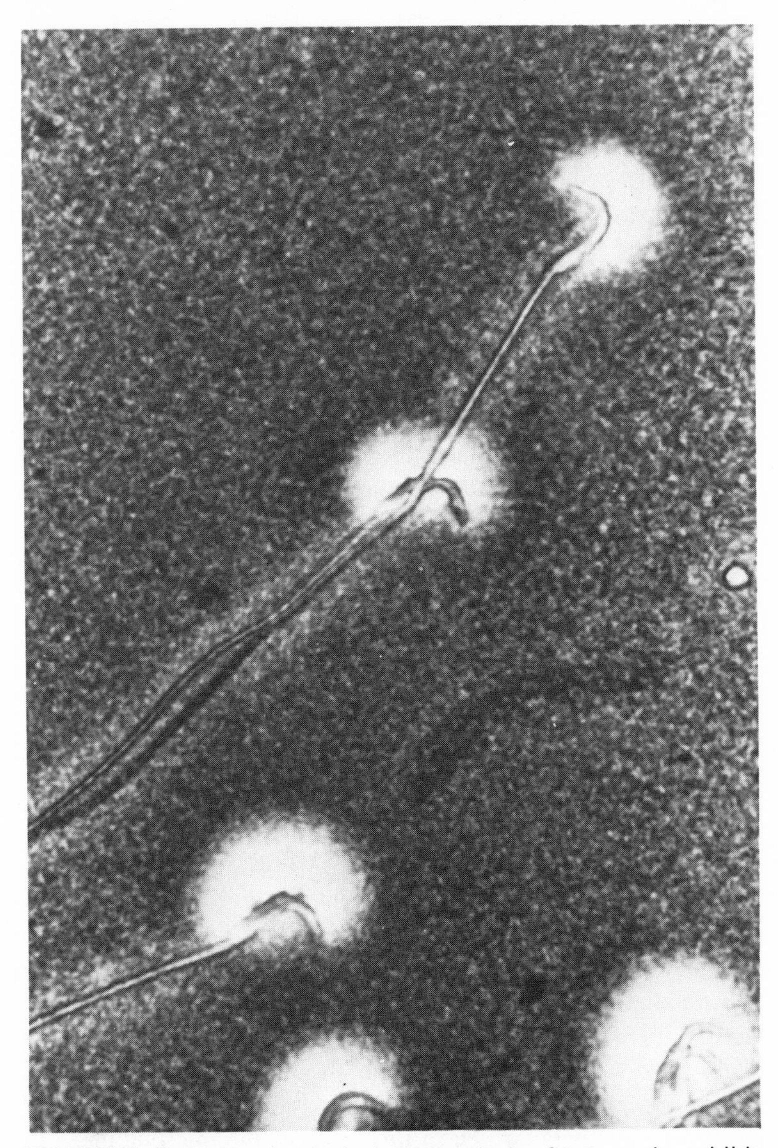

FIGURE 7. Phase-contrast photomicrograph of rat cauda epididymal sperm incubated on a gelatin film. Halos, free of silver grains, have developed around the sperm heads due to digestion of the gelatin by the acrosomal proteinase. X1000 (reduced 10% for reproduction).(Penn *et al.*, 1972.)

duct fluid (Stambaugh *et al.*, 1974). The function of these inhibitors may be to protect the egg against any soluble acrosin which may be liberated accidentally by the sperm in their vicinity (Brown and Hartree, 1976) and also to protect both the male and female genital tracts against acrosin released from degenerating sperm (Fritz *et al.*, 1975a).

Several other types of enzyme activity have been demonstrated in sperm acrosomes (Seiguer and Castro, 1972; Bryan and Unnithan, 1972, Yang and Srivastava, 1974a) or extracts of whole sperm (Srivastava *et al.*, 1970; Zaneveld and Williams, 1970; Meizel and Cotham, 1972), but their role in fertilization, if any, is at present unclear.

The perinuclear material of the spermatozoon may be organized into a pointed structure, the *perforatorium*. This is prominent in rodents (Austin and Bishop, 1958) but is not developed in some other species, e.g., man (Zamboni *et al.*, 1971). The perforatorium of rat spermatozoa has been isolated recently, and its protein composition has been determined (Olson *et al.*, 1976). It appears to consist of a single polypeptide component with a molecular weight of 13,000, of which 6.5% is cysteine. Cross-linking of nuclear histone and the perinuclear material during epididymal transit by disulfide bonds appears to provide the rigidity required for the sperm to traverse the zona pellucida (Bedford and Calvin, 1974; Olson *et al.*, 1976).

The sperm head is connected via a short neck region to the midpiece which contains a large number of elongated mitochondria spirally arranged (Fawcett, 1970). In the rat sperm the number of mitochondria has been estimated to be 500–600 (Lung, 1974). Inside the mitochondrial spiral, microtubules and dense fibers pass downward into the sperm tail. The microtubules are arranged as a central pair, surrounded by nine doublets. Dense fibers, which in turn enclose the microtubules, are generally found in animals such as mammals in which fertilization occurs internally. Their function may be to provide the additional thrust required to move the sperm through the relatively viscous fluids of the female genital tract (Nelson, 1967). The energy for this motility appears to come from the ATP generated by respiration and glycolysis. Each bull sperm contains 129×10^{-12} μmol of ATP (Mann, 1967). The contractile system of the mammalian sperm is thought to involve actin- and myosin-like proteins present in the microtubules and dense fibers, although little is known about them (Nelson, 1967). To generate ATP the sperm may utilize a wide range of nutrients. These include endogenous lipids, such as lecithin and plasmalogen, present in the midpiece, and exogenous nutrients, such as sugars, fatty acids, and amino acids, supplied by the secretions of the epididymis, the accessary glands, and the female genital tract (Mann, 1967). The sperm may utilize glycolysis immediately after insemination when a high concentration of sperm are present and switch over to respiration as the sperm concentration is reduced in the upper part of the uterus and in the oviduct (Mann, 1967).

There is now substantial evidence to indicate that the seminal plasma and epididymal secretions contain macromolecules, so-called coating antigens, that bind to the surface of spermatozoa during their maturation in the epididymis and after ejaculation (Chang, 1957; Bedford and Chang, 1962; Weil and Redenburg, 1962; Weinman and Williams, 1964; Dukelow *et al.*, 1966, 1967; Pinsker and Williams, 1967; Hunter and Nornes, 1969; Davis, 1971; Johnson and Hunter, 1972; Killian and Amann, 1973). As will be shown in Chapter 6 the removal of these coating agents appears to be the first step in sperm capacitation within the female genital tract.

Chapter 4

Gamete Transport

THE EGG

Ovulation appears to occur as a result of a thinning of the follicle wall, possibly due to proteolytic action (Espey, 1967; Parr, 1975) and an increase in ovarian contractile activity (Virutamasen *et al.*, 1976). The ovulated egg, surrounded by the cumulus oophorus, is picked up by the finger like processes (*fimbriae*) of the oviduct (Blandau, 1969). In some species, e.g., rodents, dogs, and cats, transfer of the eggs to the oviduct is aided by a capsule (*bursa*) which surrounds the ovary, except for a small aperture, and connects it to the oviduct. The bursa virtually eliminates the possibility of abdominal pregnancy by preventing eggs from accidentally escaping into the peritoneum.

The eggs are moved back and forth within the ampulla by contraction and relaxation of the muscles lining its walls. This muscular activity appears to be initiated by an as yet unidentified factor emanating from the cumulus oophorus (Blandau, 1969). Progressive movement of the eggs is brought about by the beating of the ampullary cilia and is unrelated to muscular activity. Thus, when muscular activity of the ampulla in rabbits was blocked by isoproterenol the eggs continued to be transported at the normal rate of 0.1 mm/min (Halbert *et al.*, 1976).

The number of cilia lining the wall of the ampulla is under estrogen control, as demonstrated by experiments in rabbits, in which loss of cilia followed ovariectomy and was restored by injecting estradiol (Rumery and Eddy, 1974). The amplitude of the ciliary beat may be stimulated by progesterone released from the follicle at ovulation (Borrell *et al.*, 1957).

The site of fertilization in most mammals is the ampulla. However, fertilization in the ferret may occur in the bursa and in insectivores within the follicle. Eggs must be fertilized within a relatively short period since they have a limited life span (Table 1).

The eggs are retained in the ampulla for a number of hours after fertilization by a constriction at the isthmo-ampullary junction and by a net positive pressure within the isthmus (Brundin, 1964). In rodents the small aperture in the bursa remains closed until about 24 hr after ovulation and thus allows fluid to accumulate, distending the ampulla.

On entering the isthmus, the lower portion of the oviduct which possesses fewer internal folds and fewer ciliated cells than the ampulla, a mucinous coat is secreted in some species, e.g., the opossum (Hartman, 1916), the rabbit (Gregory, 1930), and the horse (Hamilton and Day, 1945). These layers are impervious to sperm. Movement of the fertilized eggs through the isthmus usually requires approximately 2 days. During this time they undergo cleavage and are then discharged into the uterus as morulae.

The rate of forward egg movement within the isthmus appears to be controlled by the concentration of circulating estrogen, which decreases the rate of transport, and progesterone, which accelerates it (Chang, 1966). The effect of these hormones appears to be mediated by changes in the relative concentration of prostaglandins F and E within the isthmus (Saksena and Harper, 1975). These in turn may regulate contractions of the isthmus by altering β-adrenergic receptors (Rucklebusch, 1975; Hodgson and Pauerstein, 1975).

TABLE 1. Survival of Gametes in the Oviduct [a]

Species	The egg		The sperm		
	Fertilizable life (hr)	Reference	Duration of motility (hr)	Duration of fertility (hr)	Reference
Mouse	15	Marston and Chang (1964)	13	6	Merton (1939)
Rat	12	Blandau and Jordan (1941)	17	14	Soderwall and Blandau (1941)
Hamster	9–12	Yanagimachi and Chang (1961)			
Rabbit	6–8	Chang (1952)	43–50	28–36	Hammond and Asdell (1926); Seitz *et al*. (1970)
Guinea pig	20	Blandau and Young (1939)	41	21–22	Soderwall and Young (1940)
Dog	~24	Evans and Cole (1931)	268	134	Doak *et al*. (1967)
Cat			>120	50	Hamner (1973)
Pig	10	Hunter (1967)	50	24–48	Pitkjanen (1960); Hamner (1973)
Cow			96	28–50	Laing (1945); Vandeplassche and Paredis (1948)
Sheep	12–15	Dauzier and Wintenberger (1952)	48	30–48	Dauzier and Wintenberger (1952)
Rhesus monkey	23	Lewis and Hartman (1941)			
Man	6–24	Hartman (1939)	48–60	24–48	Farris (1950); Rubenstein *et al*. (1951)
Bat			149–156 Days	135–150 Days	Wimsatt (1944)

[a] See also reviews by Blandau (1969), Bishop (1969), Hamner (1973), and Ahlgren (1975).

THE SPERM

At ejaculation the sperm are propelled from the cauda epididymis and vas deferens into the urethra, where they mingle with the secretions of the accessary glands to form semen. Ejaculation may be into the vagina or directly into the uterus, depending on the species (Table 2). In some species, e.g., the dog and pig, the semen remains liquid, but in others, e.g., rodents, it coagulates (Mann, 1964). The function of coagulation may be to aid semen retention in the uterus (Blandau, 1969).

Large numbers of sperm are inseminated but relatively few reach the site of fertilization in the ampulla (Table 2). Transport is rapid, requiring only a few minutes (Howe and Black, 1963; Harper, 1973; Settlage *et al.*, 1973; Hunter and Hall, 1974). Seminal plasma constituents are transported with the sperm at least as far as the upper part of the uterus (Mann, *et al.*, 1956). Seminal plasma may also reach the oviducts, but in such a small amount that it is not detectable by the methods so far used. Sperm motility is not involved in passage of sperm up the female tract, except possibly in movement through the cervical mucus and utero-tubal junction of certain species (Blandau, 1969). Dead sperm and inert particles can be transported as rapidly as motile sperm (Bishop, 1969). The propulsive force for such transport is provided by contractions of the uterus and oviducts, possibly induced by seminal prostaglandins and other substances. In the cow sexual stimuli are known to release oxytocin which increases uterine contractions (Salisbury and Vandemark, 1961).

In some species the utero-tubal junction is known to function as a valve to regulate the number of sperm passing to the oviduct. Zamboni (1972a) perfused the reproductive tracts of female mice with gluteraldehyde at known intervals after copulation. He observed that the junction opens only briefly and then closes to exclude all but the first wave of sperm from the oviduct.

The isthmus of the oviduct appears to act as a reservoir

TABLE 2. Sperm Transport to the Site of Fertilization [a]

		Number of sperm inseminated compared with number in ampulla			Interval between insemination and appearance of sperm in oviduct	
Species	Site of insemination	Number of sperm inseminated	Number of sperm observed in ampulla	Reference	Time (min)	Reference
Mouse	Uterus	5×10^7	17	Braden and Austin (1954)	15	Lewis and Wright (1935)
Rat	Uterus	6×10^7	5–100	Austin (1948)	2–30	Warren (1938); Blandau and Money (1944)
Hamster	Uterus				2–30	Yamanaka and Soderwall (1960); Yanagimachi and Chang (1963)
Rabbit	Vagina	6×10^7	250–500	Braden (1953)	60–180	Chang (1952); Adams (1956); Braden (1953)
Guinea pig	Vagina and uterus	8×10^7	25–50	Blandau (1969)	15	Florey and Walton (1932)
Dog	Vagina				20	Evans (1933)
Pig	Vagina				15	First *et al.* (1965)
Cow	Vagina	3×10^9	Few	Salisbury and VanDemark (1961)	4–13	VanDemark and Hays (1954); Howe and Black (1963)
Sheep	Vagina	8×10^8	600–700	Braden and Austin (1954)	8–30	Starke (1949); Mattner and Braden (1963)
Man	Vagina	10^9	5–100	Doak *et al.* (1967)	5–30	Rubenstein *et al.* (1951); Settlage (1973)

[a] See also reviews by Blandau (1969) and Bishop (1969).

from which only a small number of sperm are released to the ampulla. This number appears to be regulated by the number of eggs present in the ampulla (Zamboni, 1972a; Harper, 1973), but the mechanism involved has not been elucidated.

Movement of sperm within the ampulla appears to be caused by beating of the ampullary cilia. In some regions they beat upward and in others downward. This may explain how sperm can move simultaneously in one direction and the ova in another (Blandau, 1969).

Sperm remain fertile in the oviduct for varying periods, depending on the species. In the mouse the period is about 6 hr and in women 24–48 hr (Table 1). In the bat, insemination occurs before hibernation, but ovulation and fertilization do not normally take place until the bat awakens about 150 days later.

Unused sperm in the oviduct are phagocytized, primarily by leukocytes which enter the lumen in the second half of the cycle. Surplus uterine sperm pass back to the vagina and are voided.

Wastage of male gametes is very high in mammals. Of the millions of sperm present in the ejaculate very few ever reach the egg or eggs in the oviduct. At most a few hundred or, in many cases, only a few dozen spermatozoa are ever likely to have the chance of contacting an egg. This marked dilution, together with the very small number of female gametes produced by mammals and the general lack of prolonged viability in both male and female mammalian gametes, dictates that sperm must be introduced into the female tract at or very close to the time of ovulation. This synchrony is achieved in most species by a well-defined period of estrus which restricts mating to time of ovulation.

Chapter 5

Fertilization *In Vitro*

As already indicated, recent progress in our knowledge of fertilization mechanisms is due primarily to the development of *in vitro* fertilization techniques. Despite many early claims (e.g., by Shenk in 1878; see review by Thibault in 1969), the first successful method was not devised until 1954, when Dauzier *et al.* obtained cleavage of rabbit eggs after exposing them to sperm recovered from the uterus. Since then methods have been developed for at least 11 different mammalian species, including man (Table 3).

The culture media that have been used for *in vitro* fertilization range from simple solutions, such as Krebs–Ringer bicarbonate, to complex tissue culture media, such as Medium 199 (Morgan *et al.*, 1950) and Medium F12 (Ham, 1965). These are usually supplemented with crystalline bovine serum albumin (1 to 10 mg/ml). Hoppe and Whitten (1974) have reported that fertilization of mouse eggs *in vitro* fails to occur when polyvinylpyrrolidone is substituted for albumin. The function of the albumin is not known. Possibly it may stabilize the gamete membranes, chelate the toxic ions, or stimulate the acrosome reaction (see Chapter 7). Tsunoda and Chang (1975b) have found that the presence of both lactate and glucose in the medium was essential for the penetration of a high proportion of rat eggs in

TABLE 3. Summary of Some Recent Methods of *in Vitro* Fertilization

Species	Authors	Media	Methods	Results	Comments
Rabbit	Brackett and Williams (1968) Brackett (1969)	Balanced salt solution + glucose and crystalline bovine albumin (3 mg/ml)	Uterine sperm incubated 5 hr in 4 ml medium under oil with ovulated eggs in cumulus. Fertilized eggs transferred to same medium + 10% rabbit serum and incubated 18–20 hr	50–75% of eggs cleaved, some of 8-cell. Only 25% cleaved when sperm were washed	On transplantation some normal young were born
	Ogawa *et al.* (1972)	Medium F12	Epididymal sperm incubated 5 hr in 5 ml medium with tubal eggs in cumulus. Eggs transferred to fresh medium and incubated 6 days	32% of eggs cleaved and formed blastocysts	Since sperm not washed and cumulus was present, the *in vitro* conditions were not fully defined as authors claimed
	Brackett and Oliphant (1975)	Balanced salt solution + pyruvate, glucose, and crystalline bovine albumin (3 mg/ml)	Washed ejaculated sperm capacitated by incubation in hypertonic salt solution (380 mOsm). Incubated with eggs as for Brackett and Williams (1968, 1969)	8–72% of eggs developed pronuclei or cleaved to 4-cell stage	On transplantation some normal young were born.

Rat	Toyoda and Chang (1974)	Krebs–Ringer bicarbonate solution + glucose, pyruvate, and lactate	Epididymal sperm incubated 30–35 hr in 400 μl medium with tubal eggs in cumulus	60% of eggs penetrated by 5 hr and 95% by 12–16 hr. 95% of penetrated eggs formed 2-cell embryos by 30–35 hr. Only a few formed 4-cell embryos	90% monospermic. On transplantation 21% of 2-cell embryos formed fetuses.
Mouse	Whittingham (1968)	Modified Krebs–Ringer bicarbonate solution + pyruvate lactate, and crystalline bovine albumin (1 mg/ml)	Uterine sperm incubated 5 hr in 50 μl medium under oil with tubal eggs in cumulus	10–40% formed 2-cell embryos	On transplantation a few formed fetuses
	Mukhergee and Cohen (1970)	As above	Washed uterine sperm incubated in 0.3 ml medium in cavity slide with tubal eggs in cumulus	10% formed blastocysts	On transplantation a few formed fetuses
	Toyoda *et al.* (1971)	Krebs–Ringer bicarbonate solution + glucose, pyruvate, and crystalline albumin (4 mg/ml)	Epididymal sperm incubated 5 hr under oil with tubal eggs in cumulus. Preincubation of sperm in medium for 2 hr accelerated penetration	Penetration began by 1 hr and by 2 hr 96% of eggs were penetrated	Since contribution of epididymal secretions was ignored, capacitation could not be said to have occurred in a chemically defined medium as authors claimed

TABLE 3. *(continued)*

Species	Authors	Media	Methods	Results	Comments
	Miyamoto and Chang (1972)	As above	Epididymal sperm incubated 7 hr in 0.5 ml medium under oil with tubal eggs in cumulus	64% formed 2-cell embryos	On transplantation a few formed fetuses
	Hoppe and Pitts (1973)	Modified Krebs–Ringer bicarbonate solution + pyruvate, lactate, and crystalline bovine albumin (3 mg/ml)	Epididymal sperm shaken under oil and 5% O_2 8 hr in 0.5 ml medium with tubal eggs in cumulus	92% of eggs cleaved and 88% formed morulae and blastocysts	On transplantation 37% formed viable young. Agitation and cumulus cells increased fertilization rate.
	Gwatkin *et al.* (1974)	Modified Medium 199 + crystalline bovine albumin (3 mg/ml)	Epididymal sperm capacitated in epididymal secretions or by cumulus, then incubated in 40 μl medium under oil with cumulus-free tubal eggs.	Preincubation of sperm for 2 hr resulted in 50–70% penetration in 90 min	Capacitation produced by epididymal secretions
Golden hamster	Yanagimachi (1970a)	Tyrode's solution (Tyrode, 1910)	Epididymal sperm capacitated in various blood sera and incubated 4 hr in 100 μl medium with tubal eggs in cumulus	Approximately 85% of eggs penetrated	Very variable results with this method of capacitation.

	Gwatkin and Hutchison (1971)	Modified Medium 199 + crystalline bovine albumin (3 mg/ml)	Epididymal sperm capacitated in crude β-glucuronidase and incubated 5 hr in 40 μl medium under oil with tubal eggs	50–70% of eggs penetrated	Capacitation by β-glucuronidase requires presence of epididymal secretions
	Gwatkin *et al.* (1972) Gwatkin and Andersen (1973)	As above	Epididymal sperm capacitated in dispersed cumulus and incubated 5 hr in 40 μl medium under oil with cumulus-free tubal eggs	80–100% penetrated. Completely monospermic when eggs collected 17 hr after HCG	Very reliable routine method which usually gives 100% penetration. Capacitation inhibited by glycosidase inhibitors
Chinese hamster	Pickworth and Chang (1969)	Medium F10 (Ham, 1963) + crystalline albumin (10 mg/ml)	Epididymal sperm capacitated by golden hamster cumulus, then incubated 8 hr in small drops of medium under oil with tubal eggs in cumulus	45% of eggs penetrated	
Guinea pig	Yanagimachi (1972a)	Krebs–Ringer bicarbonate solution + pyruvate, lactate, and crystalline albumin (1 mg/ml)	Epididymal sperm capacitated by incubation in medium. Tubal eggs added to 100 μl of sperm suspension	High rate of penetration after sperm were preincubated for 11–18 hr	Note long time required for capacitation

TABLE 3. (*continued*)

Species	Authors	Media	Methods	Results	Comments
Cat	Hamner *et al.* (1970)	Balanced salt solution + glucose and crystalline bovine albumin (3 mg/ml)	Uterine sperm incubated with tubal eggs in cumulus	Approximately 60% cleaved	
Cow	von Bregulla *et al.* (1974)	Ringer's solution or Medium 199 + 10% fetal calf serum	Uterine sperm incubated with ovarian eggs in cumulus	A few eggs cleaved	
Man	Edwards *et al.* (1969)	Tyrode's balanced salt solution + glucose, pyruvate, and crystalline bovine albumin (2.5 mg/ml)	Washed ejaculated sperm incubated 6–30 hr in 50 μl medium and follicular fluid with washed ovarian eggs previously matured *in vitro*	A few eggs were penetrated and formed two pronuclei	Too few eggs to permit quantitation
	Edwards *et al.* (1970)	As above and also various other media, including F10, with 20% FCS	Washed ejaculated sperm incubated in medium + ovarian eggs in cumulus for 15 hr	A few eggs were penetrated and cleaved to 16-cell embryos	Too few eggs to permit quantitation
	Soupart and Morgenstern (1973)	As above + gonadotropins	Ejaculated sperm incubated in 50 μl medium with ovarian eggs in cumulus for 24–28 hr	A few eggs penetrated	

Squirrel monkey	Gould *et al.* (1973)	Medium 199 + 20% calf serum	Ejaculated sperm incubated in 0.5–1.0 ml medium with ovarian eggs in cumulus for 24–72 hr	Approximately 50% of eggs penetrated and a few cleaved to 2-cell stage	
	Kuehl and Dukelow (1975)	Medium 199 or F10 + 20% calf serum	Ejaculated sperm incubated with *in vitro* matured eggs	Authors claim ⅓ of eggs fertilized, some cleavage	No fetuses developed on transfer to foster mothers
Dog	Mahi (1975)		Washed ejaculated sperm incubated 11–24 hr in modified Krebs–Ringer medium + crystalline bovine albumin with *in vitro* matured eggs	Sperm penetration and swelling of sperm heads in eggs	

their *in vitro* system. Calcium ions are required for the *in vitro* fertilization of mouse and rat eggs (Miyamoto and Ishibashi, 1975). The calcium appears to be essential for sperm motility and for the acrosome reaction. Media are usually adjusted to a pH of approximately 7, which is optimal for the *in vitro* fertilization of mouse eggs (Iwamatsu and Chang, 1971).

Both ovulated eggs and ovarian oocytes have been used. The latter require maturation *in vitro* but are more readily obtained than tubal eggs from some species, notably man (Lopata *et al.*, 1974). To obtain an adequate number of tubal eggs superovulation may be induced by gonadotropins. Yanagimachi and Chang (1964) found no difference in the proportion of eggs penetrated when superovulated, rather than normally ovulated, golden hamster eggs were used for *in vitro* fertilization. Spindle and Goldstein (1975) have shown that the early developmental capacity of superovulated mouse eggs is similar to that of normally ovulated gametes. The sperm may be derived by ejaculation or, more commonly, may be collected from the cauda epididymis. Rabbit sperm have been shown to develop fertility on reaching the cauda (Orgebin-Crist, 1969), and caudal sperm appear to be as fertile as ejaculated sperm (Overstreet and Bedford, 1974a). The sperm may be capacitated in the uterus, or by incubation under various conditions *in vitro* (see Chapter 6). The most consistently effective of these for hamster sperm is incubation with cumulus cells (Gwatkin *et al.*, 1972). The sperm may be capacitated prior to adding them to the eggs. Alternatively, the capacitating agent and the gametes may be incubated together for a time sufficient for both sperm capacitation and fertilization to take place. The sperm concentration is usually 1 to 5 x 10^6 per ml. This is probably a higher concentration than pertains in the oviduct, but it actually corresponds to only a few sperm in the volume occupied by an individual egg. Sperm concentrations of 6 x 10^6 to 6 x 10^7 per ml and 2 x 10^5 to 10^7 per ml have been reported to be optimal for the *in vitro* fertilization of hamster (Talbot *et al.*, 1974) and mouse (Frasser and Drury, 1975) eggs, respectively. However, in these experiments no distinction was made between the requirements for sperm capacitation and those for sperm entry into the eggs. Tsunoda and Chang (1975a) have reported that some mouse eggs are fertilized *in vitro* even when the sperm concentration is very low (less than 500 per ml). A number of factors can affect the concentration of sperm required. Among these are the adequacy of the medium, the length of time allowed for interaction of the gametes, and the initial proportion of fertile sperm in the sperm suspension. It has been suggested that only a small proportion of mammalian sperm may actually be fertile, due to errors in spermatogenesis (Cohen and McNaughton, 1974). However, the results of Tsunoda and Chang (1975a), showing that *in vitro* fertilization is possible even with very few sperm, does not support this conclusion.

To prevent evaporation and possibly also to lower the oxygen concentration to the level pertaining in the oviduct (Mastroianni and Jones, 1964; Ross and Graves, 1974), the medium is usually covered with mineral oil. The volume of medium used is typically 40 to 100 μl, although some investigators have used volumes up to 5 ml.

The results of *in vitro* fertilization may be measured in terms of the proportion of eggs penetrated (e.g., Pickworth and Chang, 1969; Gwatkin *et al.*, 1972), the proportion of eggs with two pronuclei, two polar bodies, and a sperm tail in the vitellus (e.g., Yanagimachi and Chang, 1964; Edwards *et al.*, 1969), or the proportion of eggs undergoing cleavage (e.g., Brackett and Williams, 1968; Whittingham, 1968; Hamner *et al.*, 1970; Miyamoto and Chang, 1972).

In vitro fertilization techniques vary greatly in their reproducibility, in the extent to which conditions are defined, in the degree of polyspermy, and in the extent of development obtained after sperm penetration. In some species, e.g., man, too few eggs have been fertilized *in vitro* to permit quantitation of the results. At least some systems are capable of producing normal embryos. Anderson *et al.* (1975) have shown that the ultrastructure of *in vitro* fertilized mouse eggs corresponds to that of *in vivo* fertilized eggs, at least up to the blastocyst stage. Pronuclear formation after the *in vitro* fertilization of rabbit eggs is identical to that obtained *in vivo* (Oh and Brackett, 1975). On transplantation of *in vitro* fertilized rabbit, rat, and mouse eggs to the reproductive tracts of foster mothers, a few normal fetuses or young have been obtained (see Table 3). However, the proportion of eggs capable of such development has not been determined quantitatively, and transplantation has not been attempted for most species which have been fertilized *in vitro*.

IN VITRO FERTILIZATION OF GOLDEN HAMSTER EGGS

Although *in vitro* fertilized eggs of the golden hamster have not cleaved beyond the two-cell stage and development on transplantation has not been reported at the time this book was written, the golden hamster system has been the most thoroughly studied, particularly in the author's laboratory. The system, which involves sperm capacitation by cumulus cells *in vitro*, provides 100% penetration, complete monospermy, and is probably the easiest to use and most reproducible system so far developed. For these reasons a technique for the *in vitro* fertilization of this species will be described in detail.

Female golden hamsters (*Mesocricetus auratus*), 5–7 weeks old, are superovulated to yield a mean of approximately 40 eggs per animal by administering an intraperitoneal injection of 25 I.U. pregnant mare serum gonadotropin (Gestyl, Organon Inc., West Orange, N.J.) followed 72 hr later by 40 I.U. human chorionic gonadotropin (HCG) (A.P.L., Ayerst Labs., New York). After 17 hr the animals are killed by cervical dislocation. The peritoneum is opened with scissors and the oviducts are cut away, retaining a small portion of the uterus for handling (Figure 8). The oviducts are washed with Medium 199M2 (Table 4) and then placed on the bottom of a sterile plas-

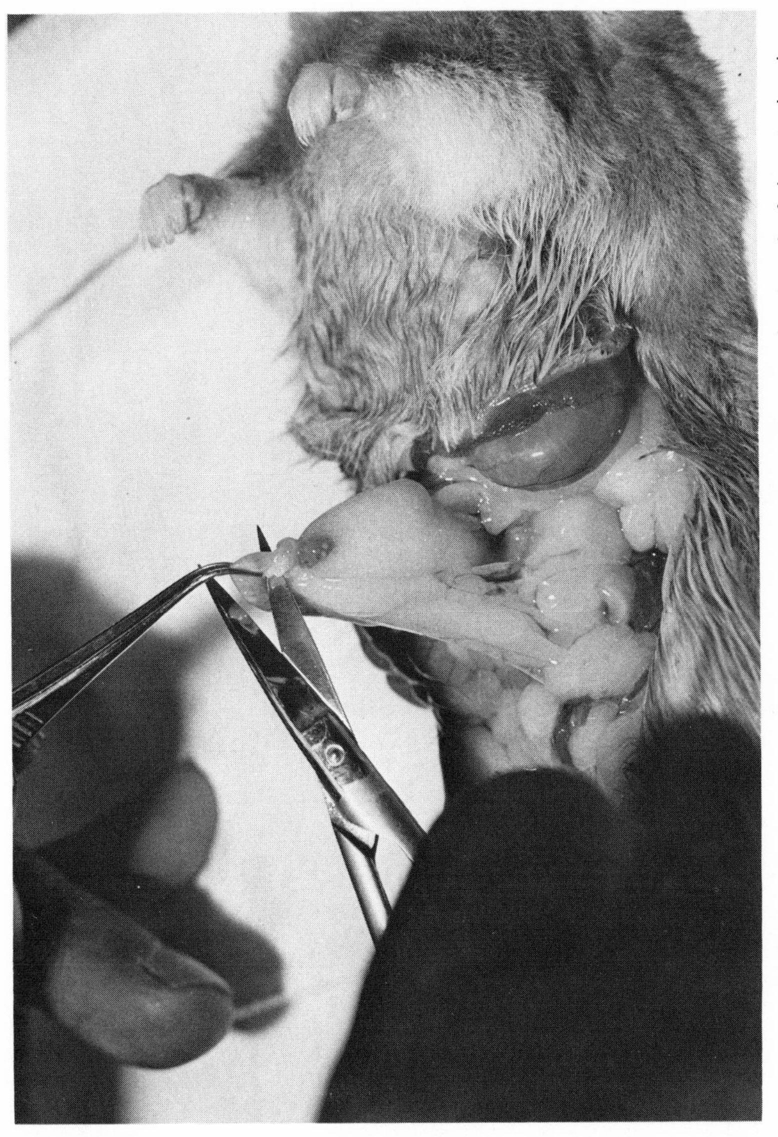

FIGURE 8. Removal of the oviduct from a golden hamster. The proximal end of the uterine horn is grasped with forceps and cut. The oviduct is then freed by cutting the ovarian bursa.

TABLE 4. Medium 199M2 for the Fertilization of Hamster and Mouse Eggs *in Vitro* [a]

Component	Concentration (mg/liter)	Component	Concentration (mg/liter)	Component	Concentration (mg/liter)
NaCl	6800.00	L-Lysine · HCl	70.00	Pyridoxine · HCl	0.025
KCl	400.00	DL-Methionine	30.00	Riboflavin	0.010
$MgSO_4$	97.00	DL-Phenylalanine	50.00	Thiamine · HCl	0.010
$NaH_2PO_4 \cdot H_2O$	140.00	L-Proline	40.00	Vitamin A acetate	0.140
$CaCl_2$	200.00	DL-Serine	50.00		
$Fe(NO_3)_3 \cdot 9H_2O$	0.72	DL-Threonine	60.00	Adenine sulfate	10.00
$NaHCO_3$	2000.00	DL-Tryptophan	20.00	ATP (disodium Salt)	1.00
		L-Tyrosine · 2Na	57.88	Adenylic acid	0.200
Sodium pyruvate	30.00	DL-Valine	50.00	Deoxyribose	0.500
Glucose	1000.00			Glutathione	0.050
Sodium acetate	50.00	α-Tocopherol phosphate	0.010	Guanine HCl (free base)	0.300
		Ascorbic acid	0.050	Hypoxanthine (sodium salt)	0.354
DL-Alanine	50.00	*d*-Biotin	0.010	Ribose	0.500
L-Arginine	70.00	Calciferol	0.110	Thymine	0.300
DL-Aspartic acid	60.00	Calcium pantothenate	0.010	Tween 80	20.00
L-Cysteine · HCl · H_2O	0.11	Cholesterol	0.200	Uracil	0.300
L-Cystine · 2HCl	26.00	Choline chloride	0.500	Xanthine (sodium salt)	0.344
DL-Glutamic Acid · H_2O	150.00	Folic acid	0.010		
L-Glutamine	100.00	*i*-Inositol	0.050	Phenol Red	20.00
Glycine	50.00	Menadione	0.010		
L-Histidine · HCl · H_2O	21.88	Niacin	0.025	K Penicillin G	100 U/ml
L-Hydroxyproline	10.00	Niacinamide	0.025	Streptomycin SO_4	100 μg/ml
DL-Isoleucine	40.00	*p*-Aminobenzoic acid	0.050	Bovine Serum Albumin	3000.00
DL-Leucine	120.00	Pyridoxal · HCl	0.025	(crystalline)	

[a] Medium is prepared by dissolving 9.87 g powdered Medium 199 (cat. No. E-11, Grand Island Biological Co., Grand Island, N.Y.) in 900 ml double glass-distilled water. The sodium pyruvate, albumin, and $NaHCO_3$ are then dissolved in the medium, and distilled water is added to make a total volume of 1 liter.

tic tissue culture dish (6 cm diameter) containing 10 ml warm paraffin oil (Saybold viscosity 125/135, Fisher Scientific Co., Pittsburgh). Before use the oil is autoclaved and shaken with a small volume of Medium 199M2, which is then allowed to separate overnight at 37°C. The equilibrated oil is stored at 37°C under 5% CO_2. This equilibration prevents shrinkage of the microculture drops.

The oviduct is held in place by grasping the uterine appendage with fine jewelers forceps (E-1947, Storz Instrument Co., St. Louis), and a tear is made in the wall of the ampulla with a second pair of forceps. The cumulus mass (Figure 9) which exudes is separated into equal portions (approximately 10 cumuli oophori), and each is pulled into a 20 μl drop of Medium 199M2, previously deposited under the oil. These operations are carried out under a dissecting microscope. Pipets and instruments are kept on warm trays prior to use (Figure 10).

Two males, 9–10 weeks of age, are killed by cervical dislocation, the peritoneum is opened, and one epididymis from each animal is deposited in a sterile plastic culture dish (3.5 cm diameter). The epididymides must be filled with sperm, as evidenced by their turgid appearance. Using iris scissors (E-3404, Storz Instrument Co., St. Louis), three cuts are made in each cauda to allow escape of the sperm. The epididymides are discarded, and 4 ml warm Medium 199M2 is added. After 1–2 min, sperm which have migrated away from the central mass are picked up with a 1-ml syringe (Figure 11). This procedure, occasionally monitored by hemocytometer count, should result in a sperm suspension which contains approximately 10^7 sperm/ml, most of which are motile. Then 20 μl of this suspension are picked up with a sterile micropipet (Corning), controlled by a tube leading to the operator's mouth and added to the drops containing the cumuli oophori (Figure 12).

The culture dish is then placed on a rocker (5–6 oscillations/min) and incubated at 37.5°C for 5 hr in an atmosphere of 5% CO_2 in air. During this period the sperm are capacitated by action of the cumulus cells (see Chapter 6 and Gwatkin *et al.*,

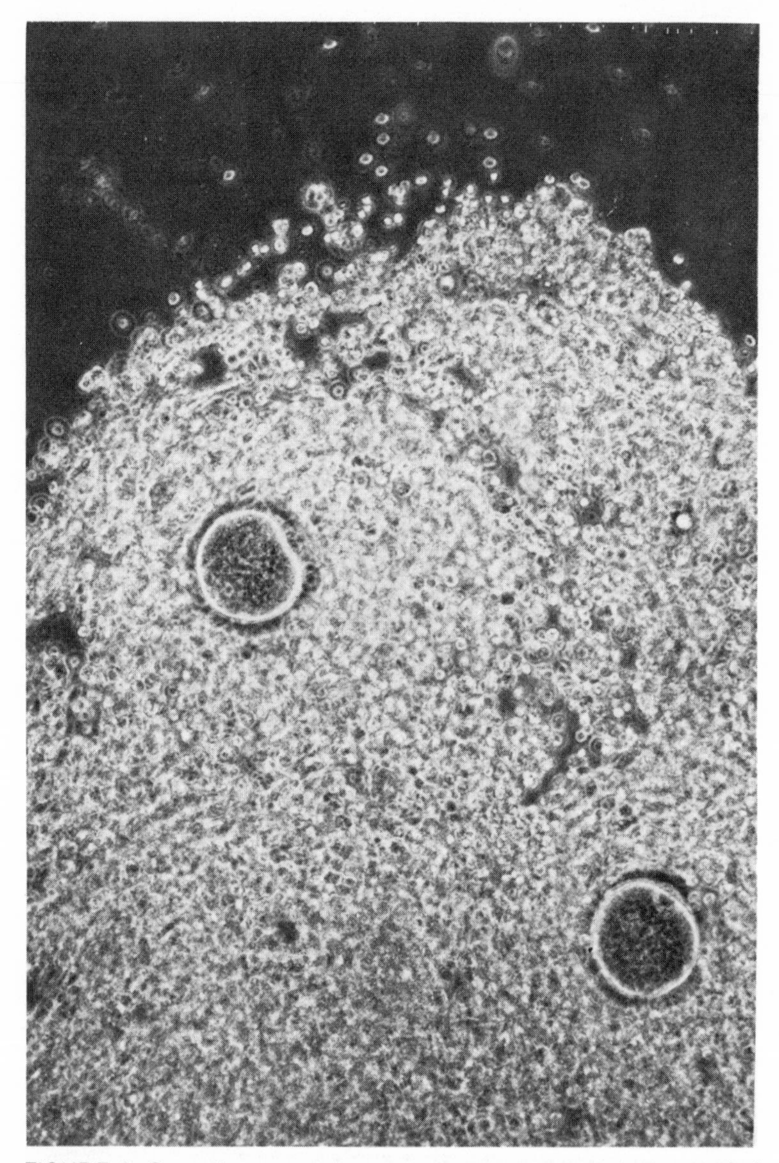

FIGURE 9. Cumuli oophori surrounding eggs, which have exuded from a cut in the ampulla of the golden hamster oviduct. X150 (reduced 10% for reproduction).

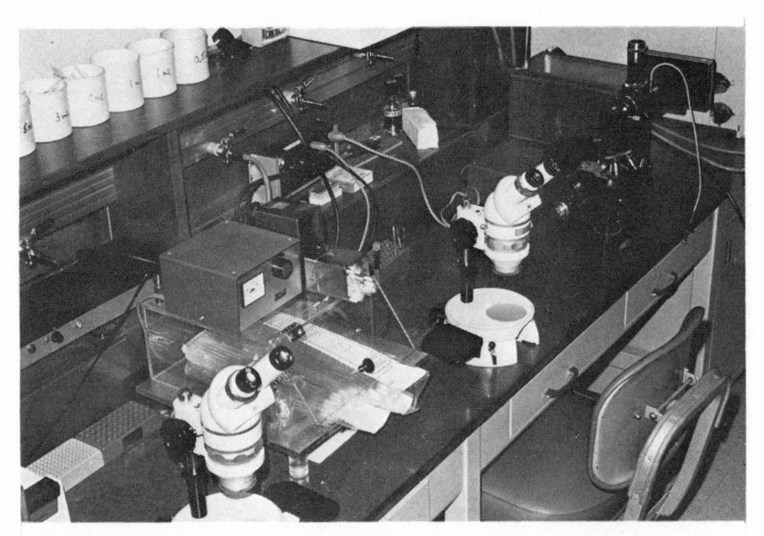

FIGURE 10. Laboratory bench arranged for *in vitro* fertilization of mammalian eggs, showing dissecting microscopes for manipulating the gametes, a phase-contrast microscope for observing and photographing sperm entry into the eggs, and a temperature-controlled platform for warming the pipets and dishes.

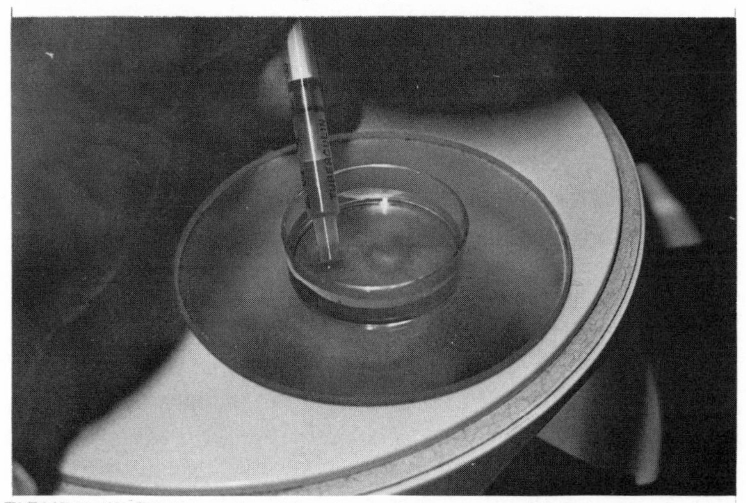

FIGURE 11. Pickup of golden hamster sperm, which have migrated from a mass of epididymal exudate placed in the center of the dish. A 1-ml syringe, without a needle, is used to obtain motile sperm which have migrated into the medium.

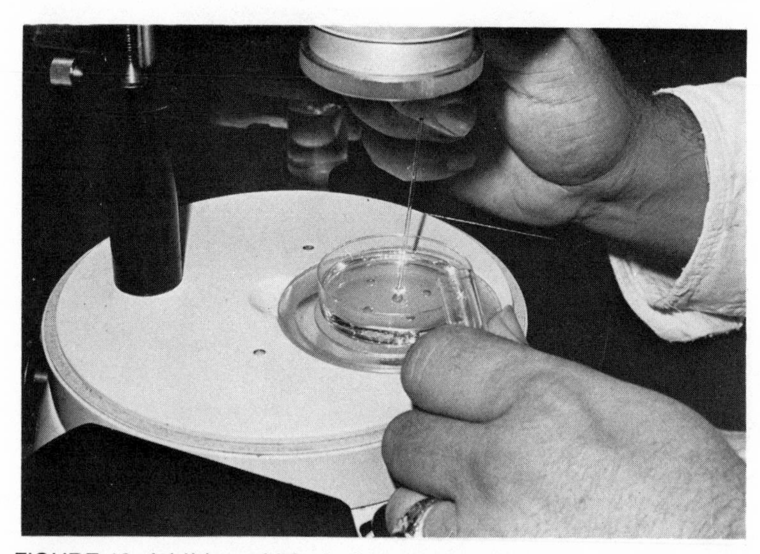

FIGURE 12. Addition of 20-μl aliquots of epididymal sperm suspension to drops containing cumuli oophori. After mixing with a pipet the dish is incubated for 5 hr on a rocker at 37.5°C to capacitate the spermatozoa.

1972). At the end of the incubation period, the culture dish is again placed on the stage of a dissecting microscope and the eggs, released by action of the sperm hyaluronidase, are removed with a finely drawn Pasteur pipet. The contents of several drop cultures are then pooled in a small tube and mixed, and a series of 40 μl drops from the pool are deposited under oil in another culture dish (6 cm diameter) using a 50-μl Corning pipet. Groups of ten cumulus-free eggs are placed in each of the drops. These eggs are obtained either from the same group of females (22 hr after HCG) or, if completely monospermic fertilization is desired, from a second group of animals injected only 17 hr previously with HCG. The cumulus is removed from the eggs by incubating them for 5 min in a hyaluronidase solution (450 U.S.P. units/ml Dulbecco's phosphate-buffered saline, supplemented with 1% polyvinylpyrrolidone). The released eggs are then washed in Medium 199M2.

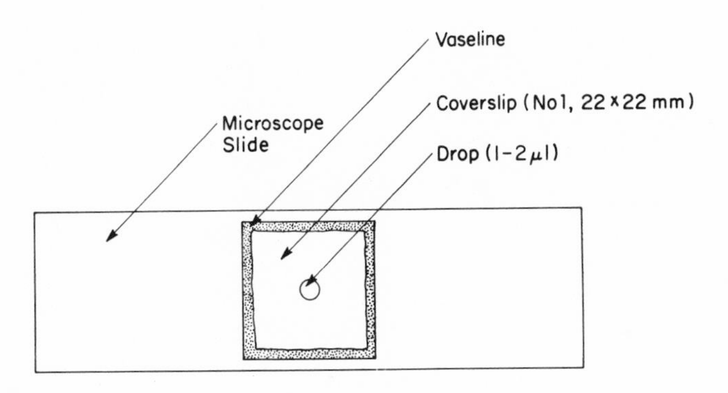

FIGURE 13. Arrangement of a slide for examination of sperm entry into eggs. The drop, containing the eggs, is compressed slightly on a layer of vaseline at the periphery of the coverslip.

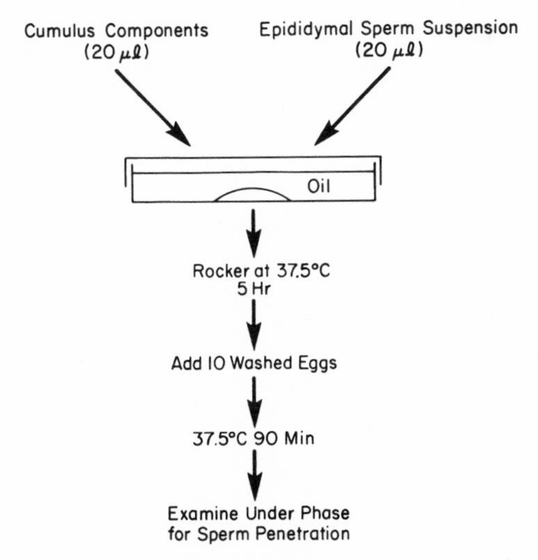

FIGURE 14. Diagram outlining steps in the capacitation of golden hamster sperm and the fertilization of eggs *in vitro*.

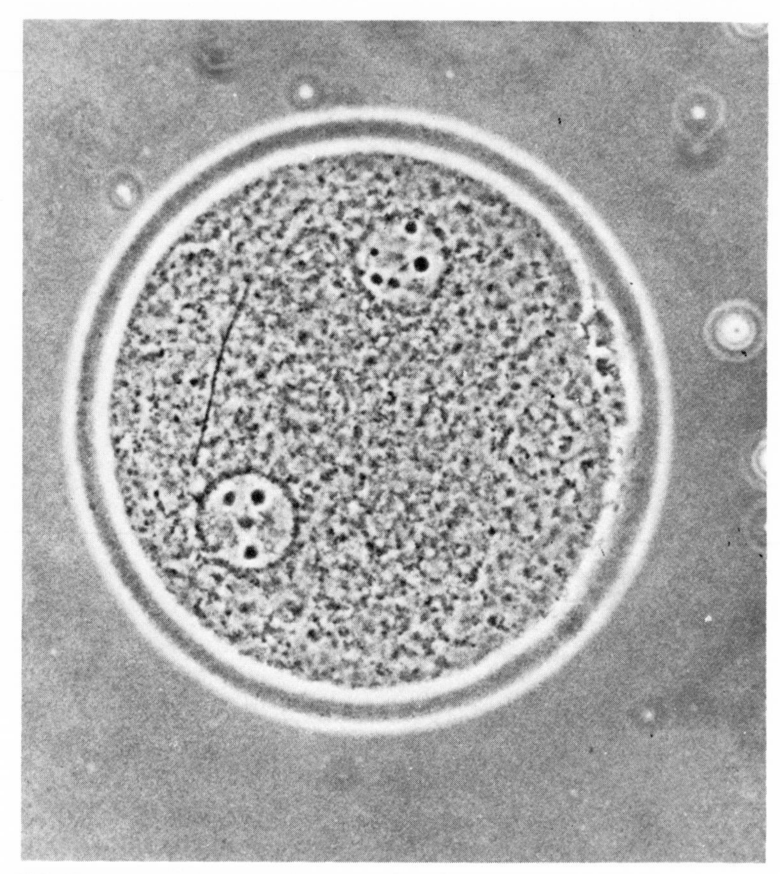

FIGURE 15. Male and female pronuclei in a golden hamster egg after incubation with capacitated sperm for 3.5 hr *in vitro*. The male pronucleus is probably the one nearest to the sperm tail, which by this time has passed completely into the vitellus. X1000 (reduced 10% for reproduction).

After incubation for 90 min at 37.5°C on the rocker, the eggs are removed and excess sperm are washed off with Medium 199M2. The eggs are then deposited on a microscope slide in 1–2 μl of medium, compressed slightly beneath a vaseline-edged coverslip (Figure 13), and examined with a phase or interference contrast microscope for the passage of sperm heads through the zona pellucida and entry into the vitellus (Frontispiece). The entire procedure is summarized in Figure 14. Typically, 100% of the eggs are penetrated. When eggs are isolated 17 hr after HCG all are monospermic. Some degree of polyspermy occurs with eggs collected 21 hr after HCG. When eggs are obtained 24 hr after HCG, all of them are penetrated by several spermatozoa. After incubation for an additional 2 hr the pronuclei appear (Figure 15).

Chapter 6

Sperm Capacitation

Before the mammalian spermatozoon can fertilize the ovum it must reside for a time in the female genital tract. This phenomenon was first described independently by Chang (1951) and by Austin (1951), who termed the process *capacitation* (Austin, 1952). The time required for capacitation to occur varies with the species (Table 5).

The first step in capacitation is probably the removal of the epididymal and seminal plasma proteins coating the sperm surface. Evidence for this comes in part from the experiments of Johnson and Hunter (1972), who showed that rabbit sperm lose their ability to bind fluorescein-conjugated antibody prepared against seminal plasma after a 10 hr incubation in the uterus. Oliphant and Brackett (1973a) prepared an antiserum against seminal plasma proteins and showed that this antiserum agglutinated ejaculated sperm but not those which had been incubated for 12 hr *in utero*. When ^{14}C-labeled antibodies were prepared against rabbit seminal plasma they bound to ejaculated sperm. On incubation of the sperm for 6 hr in uterine fluid, but not in a chemically defined medium, there was a marked decline in the binding of antibody. Experiments with similar results have been reported with boar sperm by Schill *et al.* (1975), who employed

TABLE 5. Time Required for Capacitation of Spermatozoa *in Vivo* [a]

Species	Time (hr)
Mouse	<1
Sheep	1.5
Rat	2–3
Hamster	2–4
Pig	3–6
Ferret	3.5–11.5
Rabbit	5
Rhesus monkey	5–6
Man	5–6

[a] Austin (1974).

fluorescein-conjugated antibodies against the protease inhibitors present in seminal plasma.

A second step may be an alteration in the glycoproteins of the sperm plasma membrane. Gordon *et al.* (1975a) found that the lectin, Concanavalin A, binds to the surface of rabbit sperm heads, as it also does to the tip of sea urchin sperm (Aketa, 1975). Binding of the lectin to rabbit sperm is probably not due to epididymal secretions or to seminal plasma components, since it only occurs after the sperm are washed. During capacitation *in utero* the ability to bind the lectin progressively disappears, beginning at the tip of the head (Gordon *et al.*, 1974, 1975a). Treatment of mouse and rabbit sperm with hypertonic salt solutions, a procedure which liberates protein from the sperm (Oliphant and Brackett, 1973b) and would be expected to extract peripheral proteins (Singer, 1974), induces the sperm to fertilize eggs in cumulus (Brackett and Oliphant, 1975). Pretreatment of rabbit sperm with hypertonic medium or with glycosidases was also observed to induce them to undergo an accelerated acrosome reaction (a sequel to capacitation, see Chapter 7) on transfer to follicular fluid (Oliphant, 1976). These observations suggest that peripheral proteins are being removed from the sperm plasma membrane. Such alteration of the plasma membrane would be expected to prepare it for fusion with the outer acrosomal membrane, possibly by reducing net negative charges or by increasing membrane fluidity. Consistent with this concept are the observations of Vaidya *et al.* (1971), who noted a decrease in the mobility of rabbit spermatozoa in an electrophoretic field following capacitation.

Conventional transmission electromicroscopy has failed to reveal morphological changes in the sperm plasma membrane accompanying capacitation (Bedford, 1972). However, the recent freeze-fracture studies of Koehler and Gaddum-Rosse (1975) have shown that the characteristic longitudinal strands of intramembranous particles present in fractures of the plasma membrane over the midpiece of guinea pig serum undergo dissociation during *in vitro* incubation in media that promote capacitation. These initial results suggest that by using ultrastructural techniques of greater resolution it may eventually be possible to identify the specific plasma membrane alterations responsible for capacitation.

The site of physiological capacitation of sperm within the female tract is uncertain. The initial experiments of Chang (1951, 1955) established that capacitation of rabbit spermatozoa can take place within the uterus, and this observation has been confirmed for a number of other mammalian species (see Bedford, 1970, 1972; Barros, 1974). However, there is increasing evidence that in some species at least capacitation must be completed in the oviduct (see Bedford, 1972; Barros, 1974). Insights into how this may be achieved are provided by *in vitro* experiments with hamster gametes.

Postovulatory oviduct contents containing eggs in cumulus will capacitate golden hamster sperm *in vitro* (Barros, 1968). The active component is cellular, the fluid phase separated by centrifugation being ineffective (Gwatkin *et al.*, 1972). Barros *et al.* (1967) earlier reported that "fluid" obtained from ovariectomized females would induce capacitation, but the cells and cell debris were not removed. Furthermore, such material does not simulate normal conditions within the oviduct at the time of fertilization, since it was collected 10 days after ligating both ends of the oviduct.

The cumulus oophorus by itself is a completely reliable inducer of hamster sperm capacitation *in vitro*, routinely producing sperm suspensions which fertilize all of the cumulus-free eggs added to them (Gwatkin *et al.*, 1972). In Medium 199M2 the sperm lose their motility in 2 hr, but in the presence of the cumulus oophorus rapid motility is maintained for 8 hr or more. The sperm attach to the cells (Figure 16), remain associated with them for 2–3 hr, and are then released in a capacitated state. Pretreatment of the cells with neuraminidase prevents sperm from attaching to them (Figure 16D) and blocks capacitation. Cellular microfilaments and microtubules appear to be involved in the capacitation process since preincubation of the cumulus cells with Cytochalasin B or Colcemid also blocks capacitation (Gwatkin, unpublished observation). A dialyzable factor present in the cumulus matrix is essential for cumulus cell action. Cumulus-induced capacitation occurs in a modified Medium 199, but not in Tyrode's solution (Yanagimachi, 1969a; Gwatkin, unpublished observation).

During their association with the cumulus cells the sperm are enveloped by the cumulus cell microvilli (Figure 17). Examination under the transmission electron microscope reveals that fusion between plasma membranes of the cumulus cells and the sperm does not take place, but the sperm become deeply embedded in the cells (Carter, 1974; Gwatkin and Carter, 1975). During this association the cumulus cells appear to alter the plasma membranes of the sperm by secreting glycosidases, since capacitation of golden hamster sperm by the cumulus oophorus is blocked by glucaro (1→4) lactone, a specific inhibitor of β-glucuronidase, and by 2-acetamido-2-deoxygluconolactone, a specific inhibitor of β-*N*-acetylglucosaminidase (Gwatkin and Andersen, 1973). In this connection it is of interest to note that partial capacitation of rabbit sperm can be produced by β-amylase (Kirton and Hafs, 1965) and that unwashed hamster

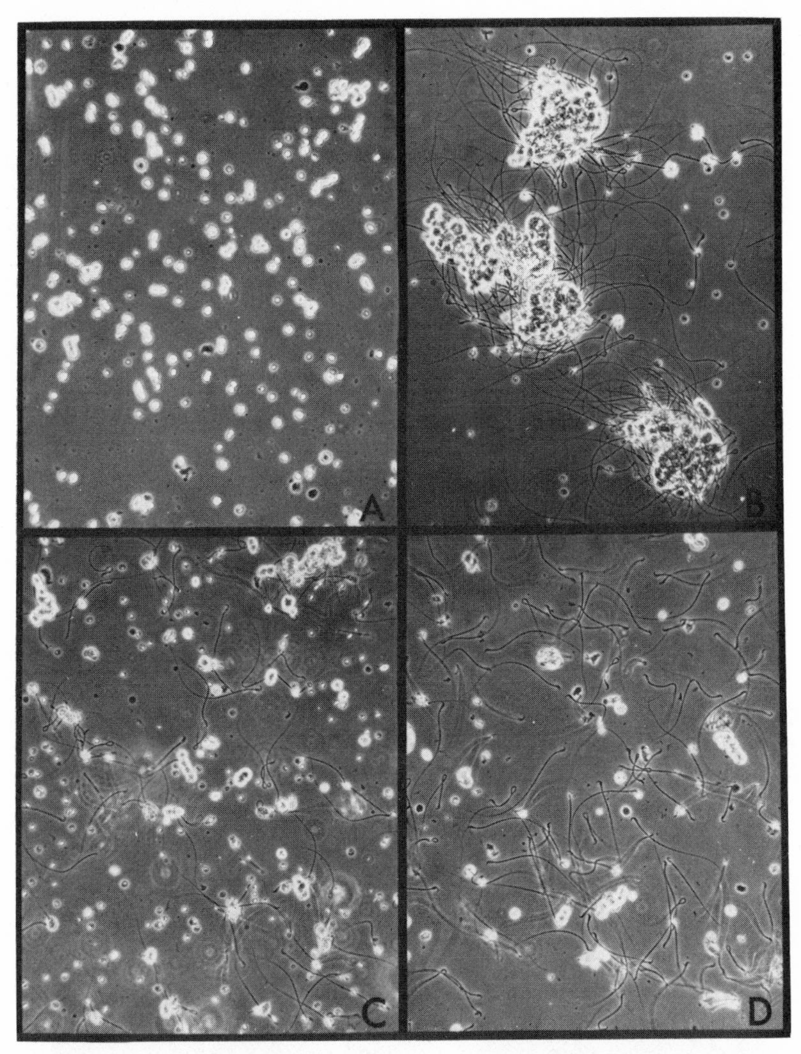

FIGURE 16. Interaction of golden hamster sperm and cumulus cells: (A) Cumulus cells dispersed with hyaluronidase; (B) 1 hr after addition of epididymal sperm, which have attached to the cells and clumped them; (C) 6 hr later, when sperm have dissociated from the cells; and (D) 1 hr after exposing neuramidase-treated cells to sperm. Note that this treatment prevents the attachment of sperm to the cells. X50. (Gwatkin *et al.*, 1972.)

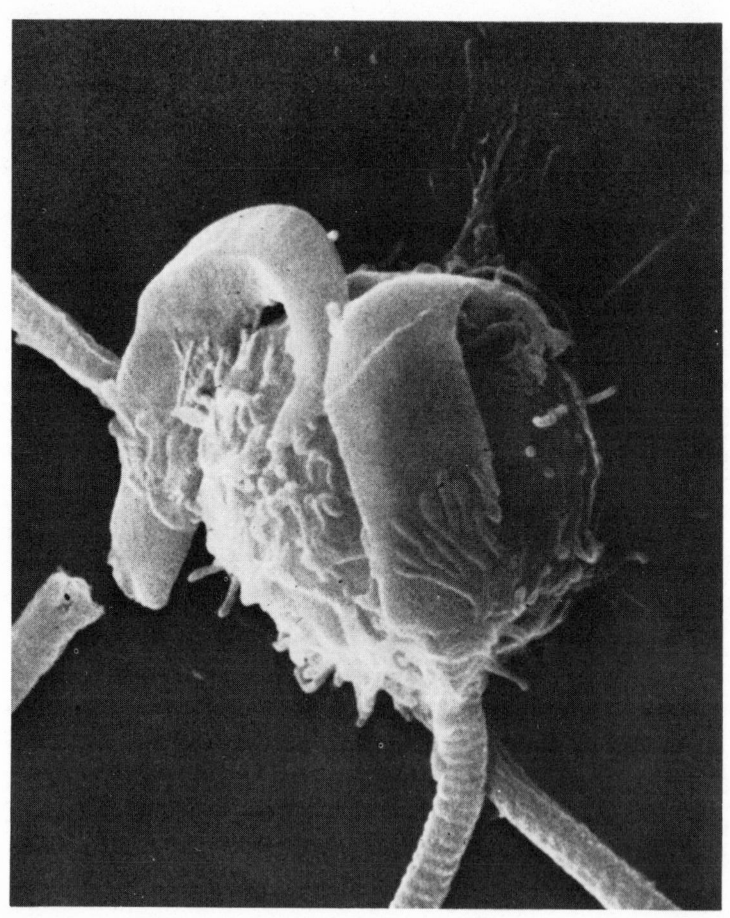

FIGURE 17. Scanning electron micrograph of cumulus cell microvilli extending around the postacrosomal region of the golden hamster sperm. X20,000 (reduced 12% for reproduction). (Gwatkin and Carter, 1975.)

sperm can be capacitated, although very slowly, by crude β-glucuronidase (Gwatkin and Hutchison, 1971). Both enzymes have been reported to remove seminal plasma proteins from sperm (Johnson and Hunter, 1972). Other evidence for a modification in the plasma membrane glycoproteins comes from the work of Gordon *et al*. (1974) who, as already mentioned, observed a progressive loss of Concanavalin A receptors over the anterior tip of the rabbit sperm plasma membrane during capacitation, suggesting the removal or alteration of carbohydrate-containing components.

It has been suggested that sperm capacitation by the cumulus oophorus may be analogous to capacitation-like processes in lower forms (Gwatkin, 1976). The spermatozoa of the hydroid, *Campanularia flexuosa*, move through epithelial passages in the gonangium to reach the eggs, and interaction of the sperm with the epithelial cells is essential before the sperm can fertilize the eggs (O'Rand, 1972). If a surface coat is removed from the cells by trypsin treatment, fertilization is prevented (O'Rand, 1974). In sponges, bed bugs, and leaches (Lord Rothschild, 1956b) sperm associate with transit cells which may function in the same way. The jelly layers surrounding frog eggs are also known to contain dialyzable and nondialyzable components which are necessary for fertilization (Elinson, 1971).

Further experiments in several different species will be required to establish the extent to which the cumulus oophorus, the oviduct, and the uterus are responsible for capacitation *in vivo*.

Hamster spermatozoa can also be capacitated *in vitro* by blood serum (Barros and Garavagno, 1970; Yanagimachi, 1970a; Morton and Bavister, 1974; Bavister and Morton, 1974) and by follicular fluid (Yanagimachi, 1969a; Gwatkin and Andersen, 1969), but these treatments do not produce sperm suspensions capable of penetrating all of the eggs added to them and they are unreliable (Mahi and Yanagimachi, 1973). Two components of follicular fluid are involved; one is dialyzable and heat stable and stimulates sperm motility, while the other is nondialyzable and heat labile and induces sperm capacitation leading to the acrosome reaction (Yanagimachi, 1969b; Morton and Bavister, 1974). The motility factor (100–200 mol. wt.), which also occurs in high concentration in the adrenal gland and to a lesser extent in the ovary, uterus, and oviduct (Bavister *et al.*, 1976), may be similar to that found in the cumulus oophorus. Rogers and Morton (1973a) have suggested that the motility factor may lower ATP and increase cyclic AMP levels within the sperm, thereby activating oxidative phosphorylation and glycolysis. This in turn would be expected to increase motility, oxygen uptake, and glucose consumption (Murdock and White, 1967)—changes associated with capacitation. The high-molecular-weight component may consist of one or more enzymes, possibly glycosidases, which would alter the sperm plasma membrane in a manner similar to the cumulus oophorus.

As a result of capacitation, hamster sperm change their pattern of motility from a progressive one to a bobbing motion of the head, apparently suited to zona penetration (Gwatkin and Andersen, 1969; Yanagimachi, 1970b). Whether this change is mediated by an alteration in the sperm surface or by some other means is not known.

Capacitation of rabbit sperm can be reversibly blocked by recoating the sperm with seminal plasma glycoproteins (Chang, 1957; Bedford and Chang, 1962; Williams *et al.*, 1967). Various glycoproteins with such "decapacitation factor" activity have been isolated, a large glycoprotein with a molecular weight in the neighborhood of 4 million (Davis, 1971) and smaller glycoproteins with molecular weights of 170,000 (Hunter and Nornes, 1969) and 115,000 (Reyes *et al.*, 1975). Capacitation of hamster sperm *in vitro* by the cumulus oophorus can be inhibited by steroids, e.g., chlormadinone and 17β-estradiol (Gwatkin and Williams, 1970; Briggs, 1973). These may stabilize the sperm plasma membrane. Sterol sulfates, which are known to be much more effective than steroids in stabilizing cell membranes (Bleau *et al.*, 1974), are likewise more effective in blocking hamster sperm capacitation by the cumulus oophorus *in vitro* (Bleau *et al.*, 1975).

Chapter 7

The Acrosome Reaction

No morphological change has been observed in mouse (Bryan, 1974) or rabbit (Bedford, 1969) spermatozoa lying free in the ampulla of the oviduct. However, as the spermatozoa pass through the cumulus oophorus the outer acrosomal membrane may invaginate to form membrane-bound vesicles within the acrosome followed by loss of the plasma membrane (Jones, 1973; Roomans and Afzelius, 1975), or multiple fusions may develop between the plasma and outer acrosomal membranes (Barros *et al.*, 1967; Franklin *et al.*, 1970; Yanagimachi and Noda, 1970b). These vesiculation processes mark the start of the acrosome reaction and probably permit the release of hyaluronidase (Figure 18b). Discharge of hyaluronidase by the sperm presumably facilitates their passage through the cumulus. Hyaluronidase release has been used as a criterion for the completion of the *in vitro* capacitation of rabbit (Lewis and Ketchell, 1972) and hamster (Rogers and Morton, 1973b) sperm and for the onset of the acrosome reaction in guinea pig sperm (Rogers and Yanagimachi, 1975a).

Vesiculation of the fertilizing sperm appears to be confined initially to a limited region of the acrosome. When capacitated sperm of the golden hamster were added to cumulus-free eggs *in vitro* the sperm penetrated the eggs, leaving behind on the sur-

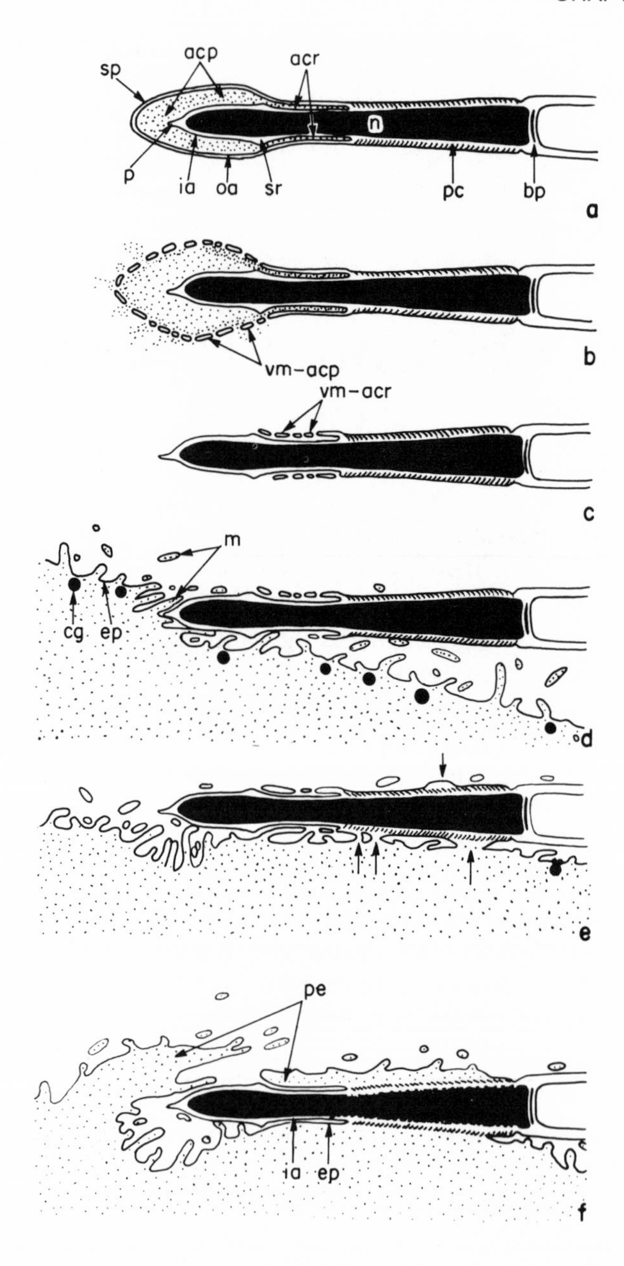
acp
acr
sp
n
p
ia
oa
sr
pc
bp
a
vm-acp
b
vm-acr
c
m
cg
ep
d
e
pe
ia
ep
f

face of the zona pellucida intact caps which had split open along their equatorial and ventral (concave) surfaces (Gwatkin *et al.*, 1975). It appears likely that vesiculation in the hamster gamete is confined to these regions, weakening them so that the acrosomal cap can separate from the sperm at the time when penetration of the zona pellucida begins (Figure 19). For a time the acrosome cap appears to remain attached to the sperm by a tether which is broken on sperm entry.

Loss of the acrosome cap probably must occur on the surface of the zona pellucida for successful fertilization to be accomplished. When hamster sperm were incubated with cumulus cells they became capacitated, i.e., capable of fertilizing cumulus-free eggs, prior to the loss of their acrosome caps, but once these were lost the sperm rapidly became infertile (Figure 20). An acrosome reaction at the zona surface is probably required to ensure that acrosin is not prematurely inactivated by the tubal environment. Loss of the acrosomal cap observed in the outer regions of the cumulus oophorus (Bedford, 1968; Yanagimachi and Noda, 1970b) may be the result of degenerative changes in supernumerary, i.e., nonfertilizing, sperm (Zamboni, 1971b).

←

FIGURE 18. Saggital sections of golden hamster sperm, showing acrosome reaction and fusion with the vitelline membrane of the egg. (Redrawn from Yanagimachi and Noda, 1970a,b.) (a) In oviduct; (b) in cumulus oophorus, showing vesiculation (start of acrosome reaction) and proposed release of hyaluronidase; (c) in zona pellucida; (d) trapped by microvilli of vitellus; (e) fusion begins between postnuclear cap (postacrosomal) region of sperm plasma membrane and the vitelline membraneof the egg; (f) fusion completed.

(acp) Acrosomal cap (anterior portion of acrosome); (acr) acrosomal collar (equatorial segment of acrosome); (bp) basal plate; (ia) inner acrosomal membrane; (m) microvilli of vitellus; (n) nucleus; (oa) outer acrosomal membrane; (pc) post nuclear cap (postacrosomal) region of plasma membrane; (sp) plasma membrane; (vm-acp) vesiculated plasma and outer acrosomal membranes of acrosomal cap (anterior portion of acrosome); (vm-acr) vesiculated plasma and outer acrosomal membranes of acrosomal collar (equatorial segment of acrosome).

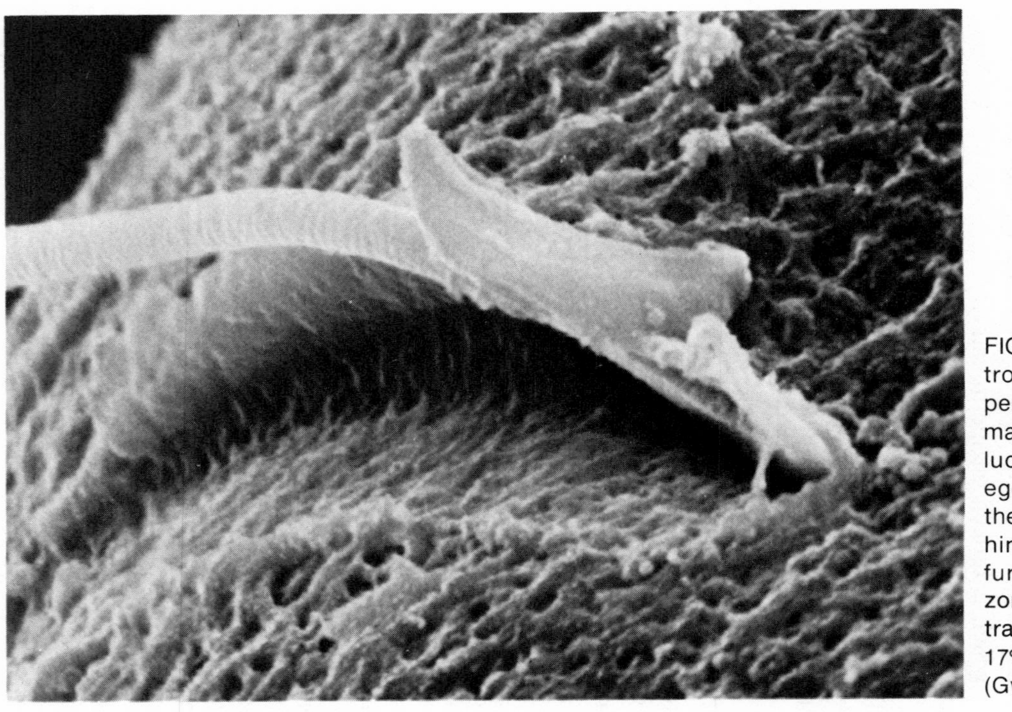

FIGURE 19. Scanning electron micrograph showing penetration of a spermatozoon into the zona pellucida of the golden hamster egg. The acrosome cap of the sperm has been left behind as the sperm plows a furrow into the sponge-like zona prior to deeper penetration. X14,000 (reduced 17% for reproduction). (Gwatkin *et al.*, 1976.)

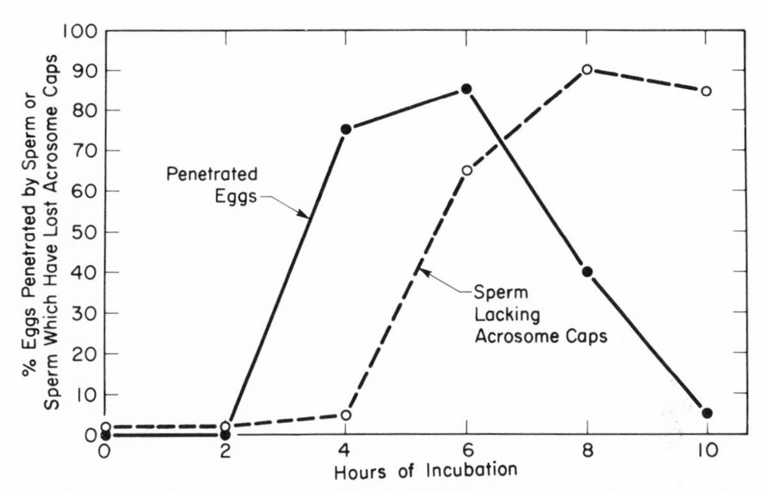

FIGURE 20. Proportion of hamster eggs penetrated by sperm and of sperm lacking acrosome caps as a function of incubation time with dispersed cumulus cells *in vitro.* (Gwatkin *et al.*, 1976.)

The acrosome reaction in golden hamster sperm appears to leave a small region of the equatorial segment of the acrosome intact. Yanagimachi and Noda (1970b) have claimed that this region contains the zona lysin and undergoes vesiculation only during passage of the sperm through the zona pellucida. However, this does not appear to be a general phenomenon, since in the rabbit this region remains unvesiculated even after the sperm has traversed the zona (Bedford, 1968), and in the mouse it appears to be lost entirely before zona penetration begins (Thompson *et al.*, 1974). Continuity of the surface membranes over the sperm head is maintained by fusion at the anterior limit of the equatorial region between the plasma membrane and the remaining fragment of the outer acrosomal membrane.

Mahi and Yanagimachi (1973) have determined some of the conditions required for the production of the hamster acrosome reaction by human serum. They found that the reaction did not occur at 16°C or at 20°C, but that at 23°C or above the time required for the reaction to take place decreased with increasing temperature up to 40°C. Above 40°C sperm survival and the proportion of sperm which underwent an acrosome reaction declined. The proportion of sperm undergoing an acrosome reaction was greatest between 230 and 343 mOsm and between pH 7.2 and 7.8.

Surprisingly, glucose, which is present in most *in vitro* fertilization media, has been found to retard the initiation of the acrosome reaction in guinea pig sperm (Rogers and Yanagimachi, 1975b). However, the presence of pyruvate and lactate appears to be required. Toyoda and Chang (1974) have observed that the omission of lactate significantly reduces the population of eggs penetrated. Miyamoto and Chang (1973b) found that the addition of pyruvate to the culture medium increased the proportion of motile mouse sperm lacking acrosomal caps and increased the proportion of eggs penetrated.

Using chromatography on Sephadex G-150 columns, Bavister and Morton (1974) found that human serum proteins with an approximate molecular weight of between 80,000 and 180,000 are the most effective in producing the acrosome reaction of hamster spermatozoa. Acrosome-reaction-inducing activity was also provided by the albumin moiety (Cohn Fraction V) of both bovine and human serum, but not by chicken ovalbumin, bovine serum, α-globulin (Cohn Fraction IV), or β-globulin (Cohn Fraction III). To be effective the active proteins required the presence of a low-molecular-weight sperm motility-stimulating factor, also present in human serum (Morton and Bavister, 1974).

Meizel and Lui (1976) have reported that the acrosome reaction of hamster sperm is inhibited *in vitro* by synthetic trypsin inhibitors and have suggested that a trypsin-like enzyme, possibly acrosin, may play a role in the acrosome reaction.

A number of reports demonstrate clearly that the presence of Ca^{2+} is required for the acrosome reaction to take place. Yanagimachi and Usui (1974) showed that the addition of Ca^{2+} (0.2 mM), but not Mg^{2+}, after guinea pig sperm were incubated for several hours in a Ca^{2+}-free medium causes an acrosome reaction to occur within 10 min. The presence of Ca^{2+} was also shown to be required for an acrosome reaction to occur following modification of the sperm plasma membrane by exposure to 0.0003% Hyamine or 0.01% Triton X-100 (Yanagimachi, 1975). Further support for the role of Ca^{2+} in the acrosome reaction is provided by the finding that compounds such as lanthanum, which block transmembrane calcium movement, inhibit the acrosome reaction of guinea pig spermatozoa (Poste, unpublished observations). Ca^{2+} also has been shown to be essential for the occurrence of the acrosome reaction in several marine invertebrates (Dan, 1967). More recently, Collins (1976) found that 1 mM ammonia or a 5 μM solution of the ionophore, A23187, which transports calcium ions across biomembranes, causes sea urchin sperm to undergo an acrosome reaction. Neither treatment is effective when Ca^{2+} is omitted from the medium.

It seems reasonable to propose that the Ca^{2+}-dependence of the acrosome reaction results from the fact that Ca^{2+} is required for membrane fusion. Ca^{2+} has been shown to play a central role in controlling numerous examples of membrane fusion in somatic cells (see Poste and Allison, 1973), and the triggering of the acrosome reaction by Ca^{2+} is reminiscent of reports in which the provision of Ca^{2+} has been shown to induce membrane fusion in germ cells (Steinhardt and Epel, 1974; Vacquier, 1975b), in diverse somatic cells (Rubin, 1970; Poste and Allison, 1973; Douglas, 1974; Thorn and Petersen, 1975; Williams and Chandler, 1975), and in model membranes of defined composition (Papahadjopoulos and Poste, 1975; Papahadjopoulos *et al.*, 1974, 1975).

By analogy with the known effects of Ca^{2+} in stimulating contractile processes and motility in somatic cells, an enhanced rate of uptake of Ca^{2+} into the sperm may also account for the marked change in the type of motility which is associated with capacitation and the onset of the acrosome reaction.

Chapter 8

Attachment and Binding of the Sperm to the Zona Pellucida

Sperm appear to bind to the surface of the zona pellucida by the plasma membrane overlying their acrosomes (Franklin *et al.*, 1970; Gwatkin *et al.*, 1976a). However, participation of the post-acrosomal region has not been excluded. Observations made by Hartmann *et al.* (1972) with a low-power light microscope have shown that hamster gametes *in vitro* associate by a complex series of interactions. During this time the sperm acrosome reaction which might explain some of their observations is probably taking place, but unfortunately, this event was not recorded by Hartmann *et al.* It will be important to repeat their experiments using the scanning electron microscope to reveal the full story of gamete association.

Summarizing the observations of Hartmann *et al.* (1972), capacitated sperm first associate loosely with the zona surface. This loose association they have termed *attachment*. It is disrupted by pipeting the gametes, occurs at 2°C or at 37°C, and is not species specific, since golden hamster spermatozoa also

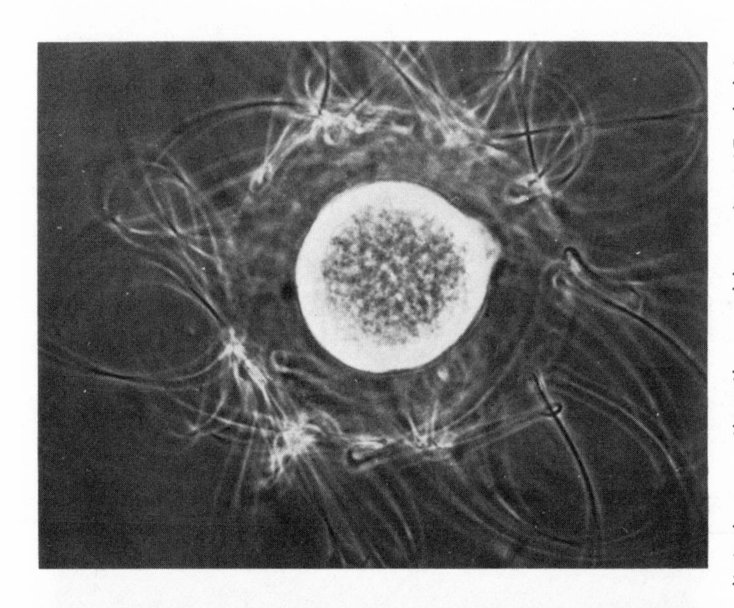

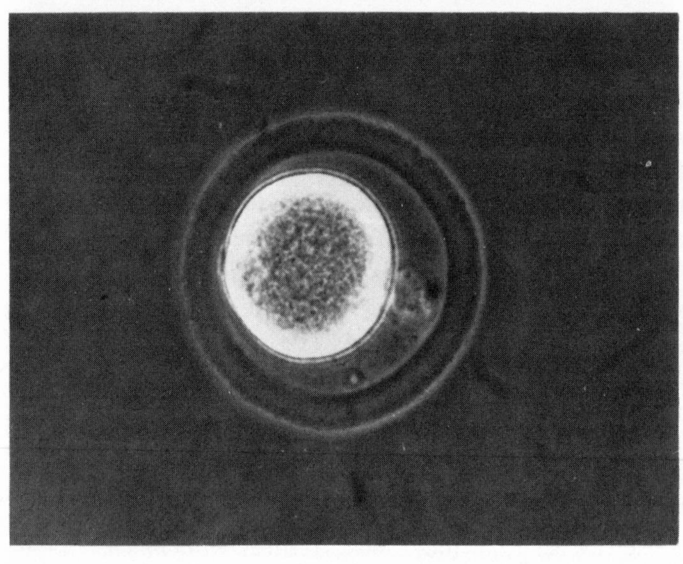

FIGURE 21. Golden hamster eggs exposed to capacitated sperm, then thoroughly washed 15 min later (left) and 40 min later (right). Note that sperm are removed at first but later remain bound. (Hartmann *et al.*, 1972.)

attach to mouse and rat eggs. After 30–40 min, a relatively tenacious union occurs, which they have termed *binding*. It is not disturbed by pipeting (Figure 21). Unlike attachment, little or no binding occurs at 2°C or to mouse and rat eggs, i.e., it appears to be relatively temperature dependent and species specific. Binding is prevented by preexposure of sperm to the trypsin–acrosin inhibitor, *p*-aminobenzamidine (Hartmann and Hutchison, 1974a). The time required for the sperm to traverse the zona and enter the vitellus is variable. Some eggs are penetrated within 5 min, but a period of approximately 20 min is required for the consistent penetration of all of the eggs added to the sperm suspension (Hartmann and Hutchison, 1974c). Yang *et al.* (1972) have reported similar *in vitro* penetration times of 4 to 20 min. Yanagimachi (1966) in a single observation recorded entry in 3–4 min. These investigators did not determine the time required for sperm binding to the surface of the zona pellucida.

ATTACHMENT OF SPERM

During attachment golden hamster spermatozoa undergo modification, possibly a conversion of proacrosin to acrosin. This was shown by transferring eggs, with sperm attached, to a second microculture, and then dislodging them by pipetting. Such sperm bind to eggs more rapidly than untreated sperm (Hartmann *et al.*, 1972), and penetration also occurs earlier (Hartmann and Hutchison, unpublished observation). Prior attachment of hamster sperm to mouse eggs does not accelerate their binding to hamster eggs (Hartmann and Hutchison, 1974a). Attachment also appears to modify the zona surface, since eggs previously subjected to attachment bind sperm rapidly (Hartmann and Hutchison, 1974c). Only 10 min of sperm attachment results in a detectable acceleration of binding, but the effect continues to develop after the attached sperm are removed. The supernatant, containing epididymal secretions, does not accelerate binding. Thus, during attachment the binding properties of both the sperm and the egg are altered.

When hamster zonae, but not those of the mouse, are included in the same microculture as the hamster gametes, binding is delayed. This was interpreted by Hartmann and Hutchison (1974b) to mean that a soluble material, designated *S1* factor, is released by the sperm attached to the zonae. It is possible that this factor may be traces of solubilized acrosin.

BINDING OF SPERM

In contrast to the binding of spermatozoa, to the eggs, which as has already been pointed out takes 30–40 min to be established, binding of sperm to the isolated zonae pellucidae requires only 5–10 min. (Figure 22). Thus, the vitellus must in some way control the normal surface interactions between gametes, preventing the rapid binding which occurs to isolated zonae. A vitelline factor was postulated to explain these results (Hartmann *et al.*, 1972). The binding of sperm to the isolated zona pellucida also seems to differ from binding to eggs in other ways. While treatment of sperm with *p*-aminobenzamidine inhibits their binding to eggs, it causes only a slight delay in their binding to isolated zonae (Hartmann and Hutchison, 1974a). Binding to isolated zonae as a function of sperm concentration follows a straight line relationship, but binding to eggs follows a sigmoidal curve (Figure 23). The latter was taken by the authors to suggest a cooperative effect, consistent with the participation of a factor from the vitellus.

It was originally supposed that normal binding might occur in two steps, the first being analogous to the rapid binding of sperm to the isolated zona pellucida. Evidence for this was the observation that some sperm bound to the eggs after incubation for 2–3 min., but not immediately thereafter (Hartmann and

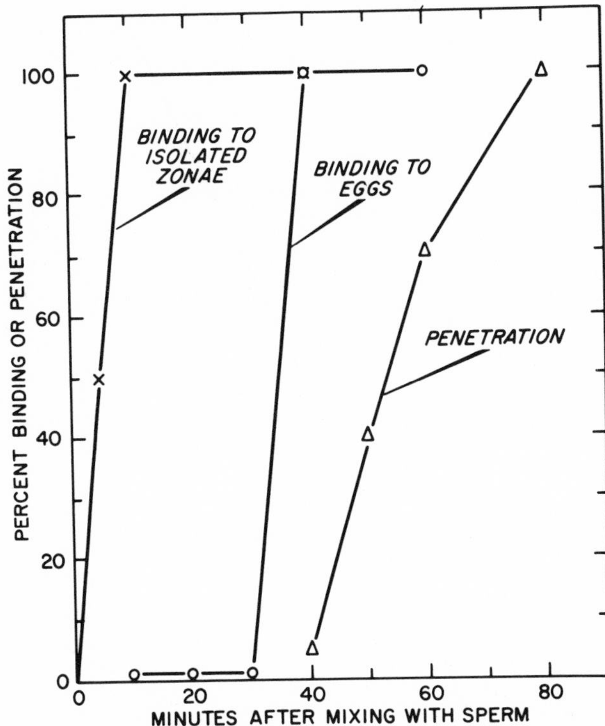

FIGURE 22. Proportion of isolated zonae pellucidae, or eggs, from golden hamster which binds sperm, and the proportion of eggs penetrated by sperm, as a function of time *in vitro*. (Hartmann *et al.*, 1972.)

Hutchison, 1974a). However, subsequent experiments showed that the number of sperm so bound was too few to account for all of those which bound at 30–40 min., and early binding was observed only at very high sperm concentrations. As a result, the authors changed their interpretation to the hypothesis that in the egg early binding, such as occurs to the isolated zona pellucida, is normally prevented (Hartmann and Hutchison, 1975). Only when an excessively high sperm concentration is applied to the eggs is this suppressive mechanism apparently overwhelmed.

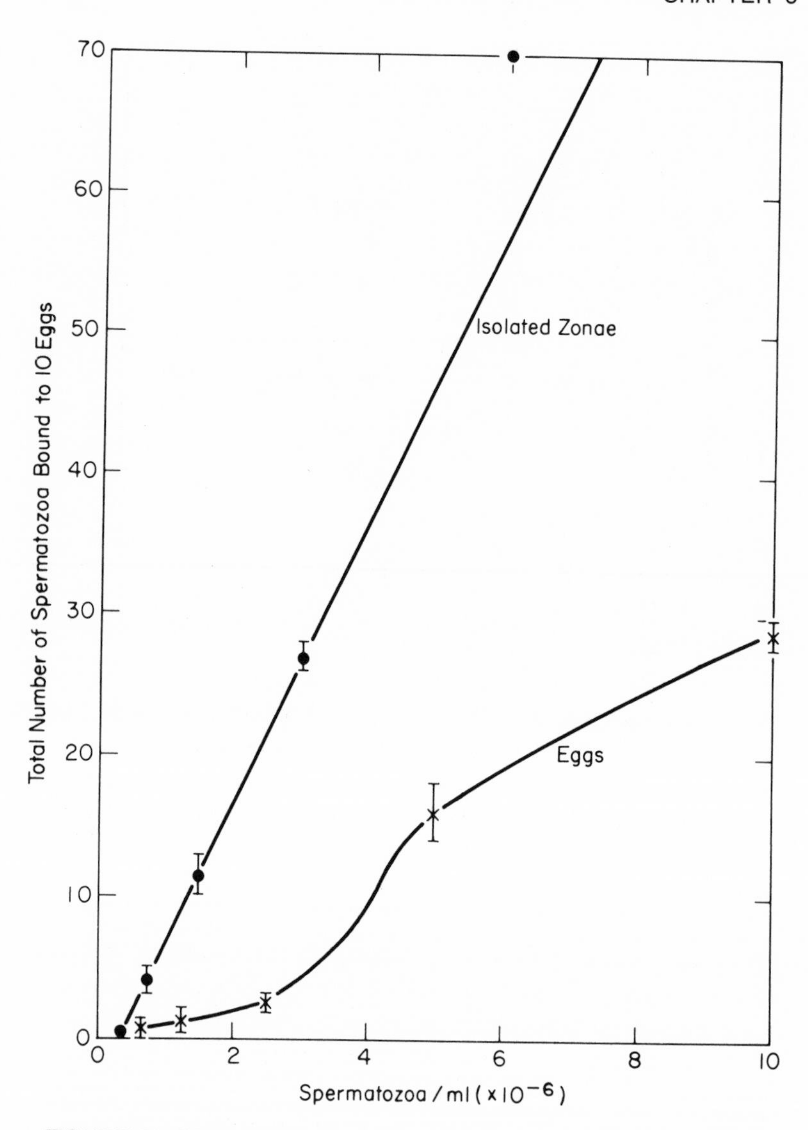

FIGURE 23. Effect of sperm concentration on the number of golden hamster sperm binding to ten eggs and to ten isolated zonae pellucidae. Note that binding to the eggs follows a sigmoidal curve, while binding to isolated zonae is linear. (Hartmann and Hutchison, 1975.)

RECEPTOR FOR SPERM IN THE ZONA PELLUCIDA

The receptor to which the capacitated sperm binds is sensitive to trypsin and chymotrypsin. Hamster eggs preexposed for 2 hr to trypsin (0.1 μg/ml) or chymotrypsin (1.0 μg/ml) no longer bind sperm, so that fertilization is prevented (Hartmann and Gwatkin, 1971). Solutions of the trypsin-like protein, acrosin, from hamster, boar, and ram spermatozoa also inactivate the receptor (Gwatkin, unpublished observation). However, binding is not sensitive to neuramidase, lysozyme, α- and β-amylase, glucoamylase, β-glucosidase, β-galactosidase, phospholipases C and D, or β-glucuronidase (Gwatkin *et al.*, 1973a). Soupart and Clewe (1965) found that pretreating rabbit eggs with partially purified neuraminidase reduced the number that were penetrated by sperm on transfer of the eggs to the oviducts of mated animals. They interpreted their results as due to the removal of sialic acid from the zona. However, their enzyme preparation may have contained a proteolytic contaminant, since it also dissolved the cumulus. In sections of ovaries sialic acid could not be detected in the zonae of mouse and rat eggs, but it did appear to be present in the zonae pellucidae of several other species, including the rabbit (Soupart and Noyes, 1964). However, the possibility that it may have migrated during histological processing was not excluded.

More recently, Overstreet and Bedford (1975) have repeated the work of Soupart and Clewe and have reported that they were unable to detect an effect of neuramidase pretreatment on subsequent sperm entry after transfer of the treated eggs to the oviducts of inseminated recipients. They also found no effect when the eggs were treated with trypsin and chymotrypsin, suggesting that the receptor-for-sperm in the rabbit zona may be different from that in the hamster.

Gwatkin and Williams (1976a,b) isolated hamster zona fragments and showed that capacitated sperm bind to the inner as well as to the outer surface of the zonae. This suggests that the receptor-for-sperm may be present throughout the zona where it would be expected to maintain close gamete association as the sperm burrows through to reach the vitellus. Although binding is relatively species specific (Hartmann *et al.*, 1972), hamster sperm were found to bind in limited numbers to the outside of mouse zona fragments (approx. two sperm per ten zonae). Binding to the inner surface, although less than to the corresponding surface of hamster zonae, was appreciable (approx. four sperm per zona). The authors suggested that the receptor on the outer surface could be partly masked by nonzona components derived from the cumulus or oviduct.

The zona pellucida of the hamster egg was found to retain its ability to bind capacitated sperm after exposure for 30 min to 60°C, a temperature close to its melting point (Gwatkin *et al.*, 1973). When 2000 zonae were incubated in 100 μl phosphate buffer solution (PBS) for 35 min at 65°C, they dissolved, and on cooling to room temperature they remained in solution (Gwatkin and Williams, 1976a,b). The addition of such zona solutions to suspensions of capacitated hamster sperm had no detectable effect on their motility but blocked their ability to bind to eggs and to fertilize them *in vitro*. A final concentration of one solubilized zona per microliter reduced the percentage of eggs penetrated by approximately 50% (Figure 24). Penetration was prevented completely by five solubilized zonae per microliter. Exposure of the eggs alone to the zona solution had no effect on subsequent penetration by sperm, showing clearly that its action was on the male gametes. Storage of hamster zona solutions for 1 week at 5°C, freezing and thawing or boiling for 5 min had no effect on their activity.

This is the first time that a soluble form of the receptor-for-sperm has been prepared from mammalian eggs. Cholewa-Stewart and Massaro (1972) observed that mouse zona melt at 68–70°C, but they did not determine the receptor activity of their zona solutions. However, a soluble sperm-binding protein has been partly purified from the vitelline membrane of sea

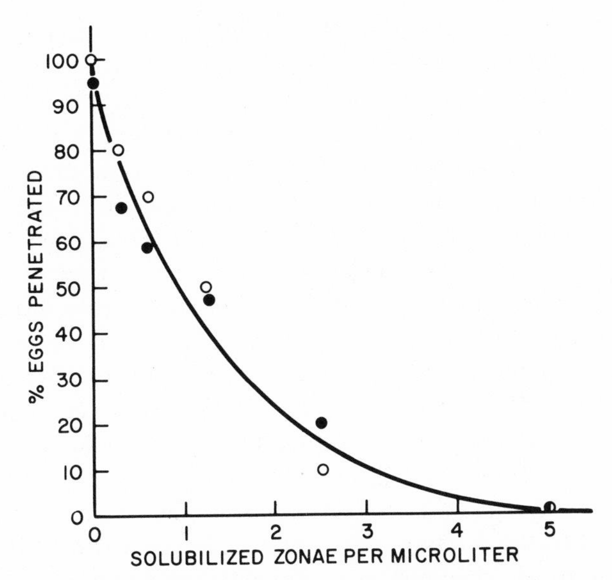

FIGURE 24. Effect of varying the concentration of heat-solubilized hamster zona pellucidae added to a capacitated sperm suspension on the proportion of eggs penetrated. Filled and unfilled circles represent two independent experiments. (Gwatkin and Williams, 1976b.)

urchin eggs (Aketa, 1967, 1973; Aketa *et al.*, 1968, 1972). To obtain the protein, the eggs are extracted with 1 M urea, the extract is dialyzed, and the active material is precipitated with calcium acetate (Aketa *et al.*, 1972). Such preparations contain 17 amino acids, several sugars, lecithin, and lysolecithin (Aketa *et al.*, 1968) and are probably very crude. Like the receptor-for-sperm in hamster eggs, the receptor in sea urchin eggs is inactivated by trypsin (Aketa *et al.*, 1972). It is also inactivated by the cleavage of S–S bonds (Aketa and Tsuzuki, 1968).

Zona solutions prepared from mouse eggs were found to produce a partial block to the fertilization of hamster eggs *in vitro*. This is in agreement with the degree of cross-binding which was noted between hamster sperm and mouse zonae. This partial inhibition suggests that the receptors in the zonae of the two species share a common determinant. As a result of the zona reaction the capacity of the solubilized zonae to prevent *in vitro* fertilization is lost (see Chapter 11).

Since the zona pellucida contains carbohydrate (Braden, 1952; Stegner and Wartenberg, 1961; Oakberg and Tyrrell, 1975) the receptor could be a glycoprotein. Several plant agglutinins (lectins), which associate with specific saccharides (Sharon and Lis, 1972), have been observed to bind to the zonae of hamster eggs (Oikawa *et al.*, 1974). One of these, wheat germ agglutinin (WGA), blocks *in vitro* fertilization, and it has been suggested that the WGA-receptor may be near, or even closely related chemically to, the receptor-for-sperm (Oikawa *et al.*, 1973; Nicolson *et al.*, 1975). However, it should be borne in mind that some lectins (Concanavalin A and phytohemogglutinin) block fertilization by inducing a cortical reaction (Gwatkin *et al.*, 1976) and that lectin-binding to zona sites unrelated to the receptor sites could mask them. Lectins could also bind to adsorbed nonzona substances derived from the follicle or oviduct (Fox and Shivers, 1975). The resistance of the receptor-for-sperm to glycosidases (Gwatkin *et al.*, 1973a) argues against a critical role for terminal sugars as receptor determinants on eggs, although they may be involved in the complementary receptor on the sperm.

Evidence that the zona pellucida is strongly antigenic is provided by the studies of Glass and Hanson (1974), who prepared antisera in rabbits against homogenates of mouse eggs in cumulus. These antisera reacted with the zonae pellucidae of the eggs and with their cumulus cells, but after absorption with mouse serum the zonae pellucidae alone reacted. Shivers and his colleagues (see review by Shivers, 1974) prepared antiserum against hamster ovary homogenates and absorbed them with hamster intestine and lung. Although not entirely specific, since cross-reaction was observed with the *theca interna* (inner cell layer of the ovarian follicle), this antiserum produced a single precipitin band against ovary extracts by double diffusion in agarose and formed a precipitate on the zona pellucida of hamster eggs (Ownby and Shivers, 1972; Dudkiewicz *et al.*, 1976), preventing their fertilization *in vitro* (Shivers *et al.*, 1972).

The experiments of Shivers' group have been confirmed for hamster eggs by Jilek and Pavlok (1975) and Oikawa and Yanagimachi (1975) and for rat eggs by Tsunoda and Chang (1976a,b). They also injected antiovary serum into females and showed that fertilization is prevented *in vivo*, thus explaining an earlier report by Shahani *et al.* (1972) that the injection of antiovary serum into mice caused them to become infertile. Oikawa and Yanagimachi (1975) have noted that in hamsters the effect lasted for 12 days, which suggests that the antibody may have coated the eggs prior to ovulation. In contrast to the earlier observations of Shivers and his colleagues (Garavagno *et al.*, 1974), who observed no reaction of hamster antiovary serum with mouse and rat eggs, Tsunoda and Chang (1976a) observed that the *in vivo* effect of their antiserum was not species specific, since hamster antiovary serum inhibited fertilization in rats and mice. They also showed that rat antiovary serum inhibited fertilization in mice and, to a lesser extent, in hamsters (Tsunoda and Chang, 1976b).

The zona antigen with which the antiovary serum reacts is present before ovulation and persists after the zona reaction and early embryonic development (Dudkiewicz *et al.*, 1975). Hamster embryos in the morula and blastocyst stages, when treated with antiovary serum and transferred to synchronized pseudopregnant recipients, failed to shed their zonae pellucidae or to implant in the uterus (Dudkiewicz *et al.*, 1975). These observations indicate that the antisera are directed against the zona as a whole and not specifically against the receptor for the sperm.

There have been no investigations on the nature of the complementary receptor on the male gamete. However, Reyes and Rosado (1975) have reported that incubation of capacitated rabbit and human spermatozoa with 1 mM *N*-4-carboxy-3-hydroxy phenylmaleimide (a specific reagent for surface-SH groups) drastically inhibits the binding of sperm to the eggs.

Chapter 9

Penetration of the Zona Pellucida by the Sperm

Evidence that acrosin is responsible for digesting a pathway for the sperm through the zona pellucida comes first from the fact that acrosin isolated from rabbit sperm will dissolve the zona pellucida of rabbit eggs (Stambaugh and Buckley, 1969). The second line of evidence is that trypsin–acrosin inhibitors inhibit fertilization both *in vitro* and *in vivo*. Stambaugh *et al*. (1969) found that ovomucoid and soybean trypsin inhibitors, each at 1 mg/ml, inhibited the *in vitro* fertilization of rabbit eggs by 36 and 90%, respectively. This corresponded to their relative effectiveness in blocking the action of rabbit acrosin on synthetic substrates. Pretreatment of capacitated rabbit sperm with the synthetic trypsin inhibitor, tosyl-L-lysyl chloroketone (TLCK), but not the chymotrypsin inhibitor, tosyl-L-phenylalanyl chloroketone (TPCK), inhibits the ability of the sperm to fertilize eggs on subsequent insemination (Zaneveld *et al*., 1970). The *in vivo* fertilization of rabbit eggs is also inhibited by incubating capacitated sperm prior to insemination with pancreatic trypsin inhibitor or partially purified rabbit seminal plasma trypsin inhibitor (Zaneveld *et al.*, 1971). Similarly the fertilization of frog eggs is blocked when sperm are incubated in naturally occurring or synthetic trypsin inhibitors prior to insemination (Greenslade *et al.*, 1973).

Although these studies tend to implicate acrosin as the zona lysin, it should be pointed out that relatively high concentrations of the inhibitors were required for fertilization inhibition and that this raises doubts about their specificity of action. In fact, Miyamoto and Chang (1973a) found that they were unable to inhibit the fertilization of hamster eggs *in vitro* with soybean trypsin inhibitor or ovomucoid. TLCK was inhibitory at 100 μg/ml, but at this concentraton it impaired sperm motility. Similar results were obtained with the chymotrypsin inhibitor, TPCK, a finding which also suggests that the inhibition was not due to a specific action on acrosin. Yet another reservation arises from the fact that in none of these studies has it been shown that the acrosome reaction was able to take place in the presence of the inhibitors. The observed failure of inhibitor-treated sperm to penetrate and fertilize eggs could simply reflect inhibition of the acrosome reaction rather than direct inhibition of the enzyme itself. Hartmann and Hutchison (1976) found that trypsin inhibitors added at the time the gametes were combined prevented the *in vitro* fertilization of hamster eggs, but when added 10 min later no inhibition occurred. They interpreted these results to mean that penetration, which normally occurs 30–40 min after the gametes are combined, may not be the result of acrosin action. While their results could mean only that insufficient time was allowed for the inhibitors to act, the observations of Hartmann and Hutchison, the failure to exclude inhibition of the acrosome reaction, the relatively high concentrations of protein inhibitors required to prevent fertilization, and their deleterious effect on sperm motility present problems to a clear demonstration that acrosin is the enzyme responsible for penetration of the sperm through the zona pellucida. While such a role seems probable, it has yet to be definitely established.

Assuming that acrosin is in fact the lytic enzyme which allows the sperm to tunnel through the zona, where is it located? Bedford (1968) has proposed that acrosin is bound to the entire inner acrosomal membrane. Yanagimachi and Noda (1970b) have suggested that it is located in the electron-dense equatorial segment of the acrosome, which in hamster sperm appears to vesiculate during the tunneling process. Evidence for the equatorial localization was suggested by Barros *et al*. (1973), who found that prolonged incubation of hamster sperm in serum, which exposed the entire inner acrosomal membrane, resulted in a loss of fertility. However, these observations do not prove that acrosin is located only in the posterior portion of the acrosome, since the medium which they employed may have inactivated the exposed enzyme. Furthermore, persistence of the posterior plasma and outer acrosomal membranes is not a general phenomenon. In the mouse they normally disappear completely before penetration begins (Thompson *et al.*, 1974). Until there is some evidence to the contrary it seems likely that the zona lysin is a peripheral protein located over most, if not all, of the inner acrosomal membrane, where it is active in a bound form (Brown and Hartree, 1976). Since soluble acrosin destroys the receptor-for-sperm in the zona pellucida (Gwatkin, unpublished observation), it is unlikely that significant amounts of free acrosin are released in the vicinity of the egg during fertilization. The function of the acrosin inhibitors found in sperm may be to inactivate any soluble acrosin released accidentally or by damaged sperm (Brown and Hartree, 1976).

Scanning electron microscopy on hamster gametes *in vitro* has shown that the spermatozoon plows a deep furrow in the zona surface (see Figure 19). The sperm then traverses the zona pellucida obliquely (Figure 25), although occasionally penetration can occur radially (Yang *et al.*, 1972). Movement through the zona appears to be facilitated by the bobbing movement of the head which develops following capacitation (see Chapter 6) and presumably brings the acrosin on the inner acrosomal membrane into close association with zona material. In rodents the

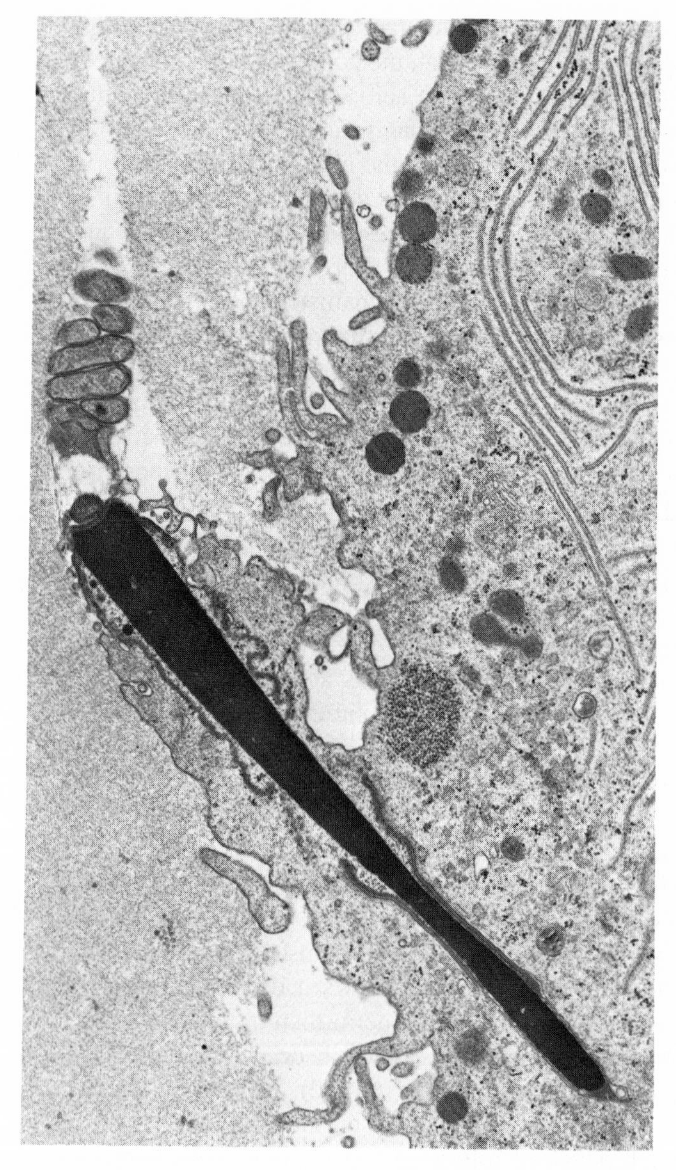

FIGURE 25. Transmission electron micrograph showing penetration of the hamster egg by a spermatozoon. A sharply defined oblique channel has been made by the sperm in the zona pellucida (top of photograph), and the vitellus of the egg has formed a fertilization cone around the sperm head, which lacks its acrosome cap and is beginning to lose its nuclear membranes. An aggregate of small particles has formed near the fertilization cone and the cortical granules have disappeared from the site of sperm entry. X20,000 (reduced 10% for reproduction).

process may also be aided by organization of the perinuclear material into a relatively rigid structure, the perforatorium (Yanagimachi and Noda, 1970b).

The time required for the sperm to traverse the zona pellucida appears to be variable. Yang *et al*. (1972) removed cumuli oophori containing eggs and sperm from the hamster oviduct and observed that 4 to 22 min was required for zona penetration. After traversing the perivitelline space, a rapid process requiring only 1–2 sec (Yanagimachi, 1966), the sperm moved some distance away from its point of entry before fusing with the vitelline membrane. This movement caused rotation of the vitellus.

Sperm can penetrate the cumulus oophorus and zona pellucida of immature oocytes at the germinal vesicle stage, as shown by Overstreet and Bedford (1974b) using rabbit gametes. However, they do not fuse with the vitelline membrane. Similar results with mouse gametes were reported by Iwamatsu and Chang (1972). Penetration of the zona pellucida is species specific, e.g. capacitated hamster sperm do not enter mouse (Yanagimachi, 1964) or rat (Barros, 1968) eggs. This is due, at least in part, to the fact that limited binding occurs with the zona pellucida of these species (Hartmann *et al.*, 1972). It has also been observed that capacitated guinea pig sperm do not penetrate hamster eggs, although they will fuse with zona-free eggs (Yanagimachi, 1972b).

Penetration of the sperm through the zona pellucida of the rat egg *in vitro* shows a sharp optimum at pH 7.7, but the optima for mouse (pH 7.3–7.7) and hamster (pH 6.8–8.2) sperm are relatively broad (Miyamoto *et al.*, 1974).

Chapter 10

Fusion of the Sperm with the Vitellus

Studies with hamster gametes *in vitro* have shown that the microvilli of the vitellus wrap around the head of the sperm and fuse with its plasma membrane in the postacrosomal region (Yanagimachi and Noda, 1972). The role of the postacrosomal plasma membrane in initiating fusion between mammalian gametes is in striking contrast to gamete fusion in marine invertebrates in which initial contact with the egg membranes is made by an apical process extending from the inner acrosomal membrane (Colwin and Colwin, 1967; Epel, 1975; Summers *et al.*, 1975).

The importance of the microvilli, commonly found in cell fusions induced by viruses and other agents (Poste, 1970), is that their low radius of curvature may facilitate fusion by overcoming the net negative charges on cell surfaces (Pethica, 1961; Poste, 1970). Few microvilli have been observed over the second metaphase spindle of mouse eggs, and this could account for the fact that gamete fusion is rare in this region (Johnson *et al.*, 1975). In contrast to the sperm plasma membrane, Nicolson *et al.* (1975) have observed a uniformly high degree of lectin receptor mobility throughout the oolemma. This observation suggests that fusion can potentially take place at any point on the egg surface, provided the correct apposition of the post-acrosomal region of the sperm and the vitelline surface has first occurred. An additional function of the egg microvilli may be to ensure this correct orientation.

Following the initial fusion between the plasma membrane of the egg and the plasma membrane of the postacrosomal region of the spermatozoon, tongues of egg cytoplasm surround the anterior portion of the sperm head (see Figure 18f). This process has been observed by transmission electron microscopy during the fertilization of rat (Pikó, 1969), rabbit (Bedford, 1970), and hamster (Yanagimachi and Noda, 1970b) eggs. However, no fusion occurs between the inner acrosomal membrane and the vitelline membrane, so that for a time a vacuole persists in this region (Pikó, 1969).

During fusion, the gamete membranes intermix. This has been demonstrated by Yanagimachi *et al.* (1973), who used colloidal iron (Gasic *et al.*, 1968) to label the acidic ionic groups on hamster gametes undergoing fertilization *in vitro*. Before fusion occurred the vitelline membrane of the egg, but not the plasma membrane of the sperm, became heavily labeled. As fusion began the label appeared over the sperm head, indicating intermingling of the sperm and egg membrane components. The conditions for this fusion reaction have not yet been established, but Frye and Eddidin (1970) have observed that the intermixing of H-2 antigens in mouse and human cell lines, fused with Sendai virus, is temperature sensitive but requires neither energy nor protein synthesis. It would be of interest to carry out similar experiments with mammalian gametes, labeled with fluorescein. Immunological markers might also be employed, since fusion can occur between the sperm and vitelli of different species (see below).

As the sperm fuses with the vitellus the egg cytoplasm rises up to form a protruding region, the *fertilization cone* (see Figure 25). In the sea urchin, microfilaments have been observed paralleling the long axis of the cone, and it has been suggested that they may be involved in its formation and later resorption (Longo, 1973).

Gamete fusion seems to be possible only after the plasma membrane of the sperm has been modified by capacitation. Yanagimachi and Noda (1970c) showed that golden hamster sperm, which previously had been capacitated by incubation for 3 hr in 50% bovine follicular fluid, fused with eggs from which the zonae pellucidae had been removed with 0.1% trypsin. However, if they were incubated for only 10 min, fusion did not occur. Niwa and Chang (1975) found that rat sperm required incubation for 3–4 hr before they could fuse. Capacitated sperm will fuse with the vitelline membrane of immature (germinal vesicle) eggs, but the process is less efficient than with eggs which have undergone meiotic maturation or are in the process of doing so. Barros and Munoz (1973) found that only 7% of zona-free hamster oocytes, removed from the ovary immediately after an injection of HCG, incorporated capacitated sperm, but 6 hr later, when the oocytes had matured to metaphase I, 90% incorporated male gametes. Iwamatsu and Chang (1972) reported that only a few mouse oocytes at the germinal vesicle stage were penetrated by sperm, but that this proportion increased with the degree of maturation.

Fusion of the sperm with the vitellus is relatively nonspecies specific, when compared to zona penetration. Hanada and Chang (1972, 1976) found that capacitated mouse sperm will enter rat vitelli and that mouse and rat spermatozoa will penetrate hamster and rat vitelli. As with homologous gamete fusion, capacitation of the sperm is required (Hanada and Chang, 1976).

Zamboni *et al*. (1972) have reported that cumulus cells may migrate through the zona pellucida of human eggs *in vitro* and become enclosed in ooplasmic vacuoles. However, they do not fuse with the oolemma.

Chapter 11

The Prevention of Polyspermy

THE CORTICAL REACTION

Fusion of the cortical granules with the oolemma and a discharge of their contents into the perivitelline space occurs when the sperm reaches the vitellus (Szollosi, 1967; Pikó, 1969). Studies with hamster gametes *in vitro* have shown that the trigger for this so-called *cortical reaction* is fusion between the gamete membranes and that simple contact is not sufficient (Gwatkin *et al.*, 1976b). Penetration of the sperm into the vitellus is not needed, since it was found that capacitated spermatozoa would induce the reaction even after they were frozen and thawed so as to prevent penetration. The cortical reaction is propagated around the egg from the point of association of the fertilizing sperm with the vitelline membrane (Braden *et al.*, 1954).

A variety of treatments have been employed to artificially induce the cortical reaction of hamster eggs (Gwatkin *et al.*, 1976b). These include electrical stimulation (Figure 26) and exposure to positively charged microbeads, neuraminidase, boro-

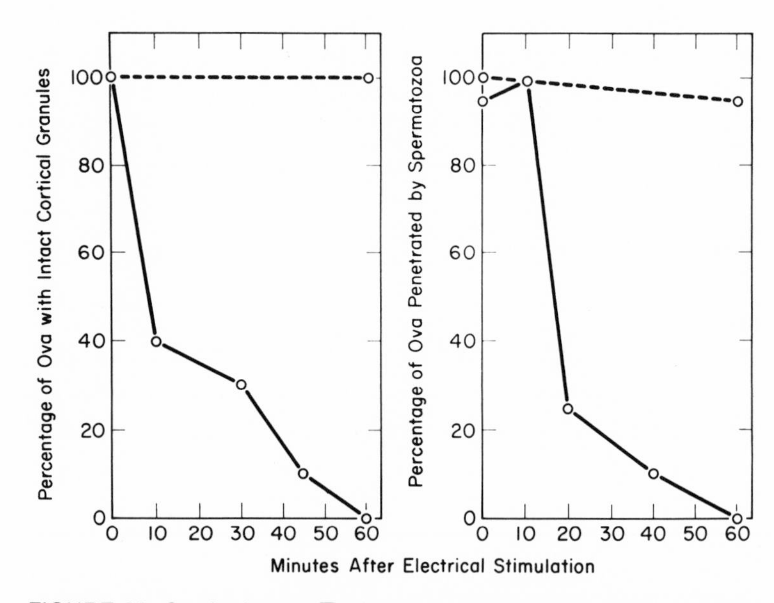

FIGURE 26. Cortical granule discharge and development of a zona reaction, induced by the electrical stimulation of golden hamster eggs. Dotted lines represent control (unstimulated) eggs. (Gwatkin *et al.*, 1973.)

mycin, 1, 3-*bis*-(4-chlorocinnamylideneamino) guanidine, Concanavalin A, and phytohemagglutinin. Action of these agents may be explained by assuming that the cortical granules, like other intracellular organelles (Poste and Allison, 1973) carry a net negative charge, which would normally keep them out of contact with the negatively charged inner surface of the vitelline membrane (Steinhardt *et al.*, 1971). Attachment of a positively charged bead to the oolemma would be expected to disturb its polarity at this point and set off a wave of depolarization. Steinhardt *et al.* (1971) have recorded such a depolarization following the fertilization of sea urchin eggs. The resulting loss of electrostatic repulsion between the cortical granules and the vitelline membrane might then reduce the distance between them to the 10 Å required for membrane fusion (Poste and Allison, 1973). Perhaps an important function of the zona pellucida is to protect the vitellus from positively charged particles in the oviduct before fertilization. Neuraminidase is known to reduce the net negative charge on the outside of cell membranes by removing terminal sialic acid residues (Weiss, 1969) and this also would cause membrane depolarization. Lectins, such as Concanavalin A and phytohemagglutinin, are thought to crosslink the integral proteins of cell membranes, pulling them into clusters (Singer and Nicolson, 1972). This reorganization might alter the charge on the inside of the membrane via transmembrane linkages (Ji and Nicolson, 1974). Boromycin (Fig. 27) is a monovalent ionophore which binds to cell membranes and specifically induces loss of K^+ from the cell (Pache, 1975). The effect of the hamster egg would probably be to produce depolarization of the oolemma. Biguanides and guanidines are known to bind to membrane lipids, altering the potential of the membrane (Schaeffer *et al.*, 1974).

Studies by Steinhardt *et al.* (1974) have shown that the cortical reaction can also be induced in sea urchin, toad, and hamster eggs with the ionophore, A23187 (Figure 28). For the hamster gametes a 2-min exposure to a 3 μM solution is sufficient to block fertilization. Unlike boromycin, which transports monovalent cations, A23187 is a transporter of divalent cations, such as Ca^{2+} (Pressman, 1973; Schaeffer *et al.*, 1974). Since A23187 induced the cortical reaction in both sea urchin and hamster eggs in the absence of Ca^{2+} or Mg^{2+} in the medium, Steinhardt *et al.* (1974) have suggested that A23187 may act by releasing a divalent cation from intracellular stores rather than by transporting divalent cations into the eggs. This is probably Ca^{2+}, since the bulk of the Mg^{2+} in the unfertilized sea urchin egg is in a free form, whereas the bulk of the Ca^{2+} is bound (Steinhardt and Epel, 1974). An increase in dialyzable (i.e., free) calcium has in fact been recorded following the fertilization of sea urchin eggs (Nakamura and Yasumasu, 1974). Since binding of Ca^{2+} to one side of a phosphatidylserine bilayer has been observed to alter its polarity (Wobschall and Ohki, 1973), A23187 may also be inducing the cortical reaction through depolariza-

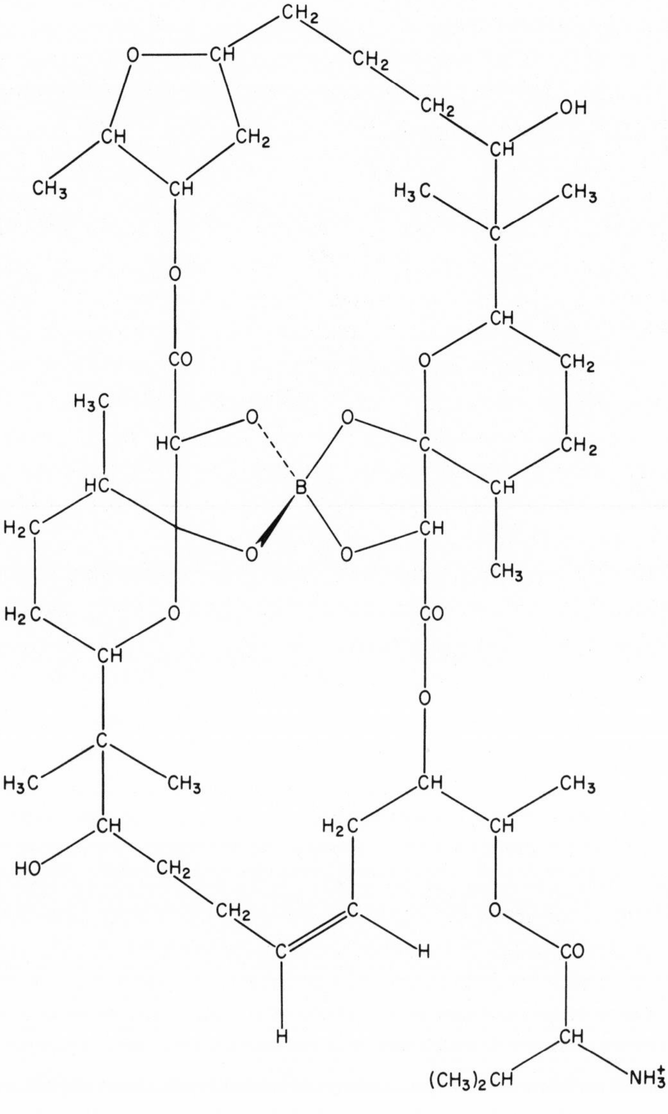

BOROMYCIN ($C_{45}H_{74}BNO_{15}$)

FIGURE 27. Structure of the monovalent ionophore, Boromycin. (Pache, 1975.)

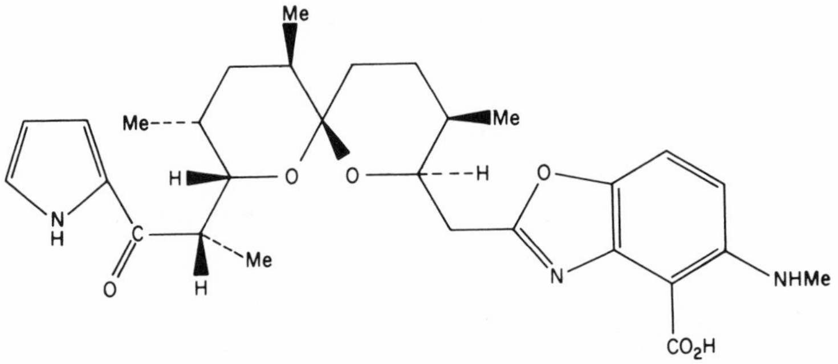

FIGURE 28. Structure of the divalent ionophore, A23187. (Chaney *et al.*, 1974.)

tion of the hamster egg membrane. It is also possible that Ca^{2+} release may affect the cortical granules directly. Vacquier (1975b) has obtained sea urchin oolemma preparations with attached cortical granules by sticking eggs onto a layer of protamine and then shearing off the unattached membranes and egg cytoplasm. When he added Ca^{2+} to such preparations the granules discharged in a propagated wave. However, he did not observe fusion with the oolemma, so that this response may not be equivalent to a true cortical reaction (Epel and Johnson, 1976).

THE VITELLINE REACTION

As a result of the cortical reaction much of the original vitelline membrane is reconstituted from the membranes of the cortical granules (Graham, 1974). This change, the *vitelline reaction*, prevents sperm entry and is the primary block to polyspermy in the few mammalian species which lack a zona reaction. It is also possible that the cortical granule products may react with the oolemma. However, the relatively slow onset of the vitelline block (see below) suggests that it is not produced by this means but by membrane reformation. Reformation of the membrane would be expected to alter its properties. Following fertilization the rabbit vitelline membrane binds more Concanavalin A (Gordon *et al.*, 1975b), and vitelli become agglutinable by lower concentrations of this lectin (Pienkowski, 1974). The number of negatively charged groups on the rabbit vitelline membrane increases after fertilization (Cooper and Bedford, 1971). These groups appear to be contributed by sialic acids in the *N*-acetyl-*O*-diacetyl configuration, since their removal requires saponification prior to neuraminidase treatment.

Resistance of the vitelline membrane to sperm entry appears to develop more slowly than the zona reaction. Barros and Yanagimachi (1972) have estimated that in the hamster egg 2 to 3.5 hr are needed for the vitelline block to develop compared with less than 15 min for the zona reaction to become effective. The vitelline reaction may be inhibited by the injection of chelating agents into the oviducts of rats prior to mating (Pikó, 1961; von der Borch, 1967). Possibly such treatment may prevent membrane reorganization by sequestering Ca^{2+}.

THE ZONA REACTION

The concept that the products of cortical granule discharge alter the zona pellucida so that sperm can no longer penetrate it was originally proposed by Austin and Braden in 1956. In a few mammalian species, e.g., the rabbit (Braden *et al.*, 1954), this reaction is absent.

Proof that the zona reaction is produced by the cortical granule contents was provided by Barros and Yanagimachi (1971), who collected material discharged from fertilized hamster vitelli and found that eggs pretreated with it became infertile. Gwatkin *et al.* (1973b) found that cortical granule material, collected either from fertilized or electrically pulsed vitelli (Figure 29), altered the zonae pellucidae of eggs so that sperm did not bind to them and fertilization was prevented. The material is heat labile (Gwatkin and Williams, 1974), effective on

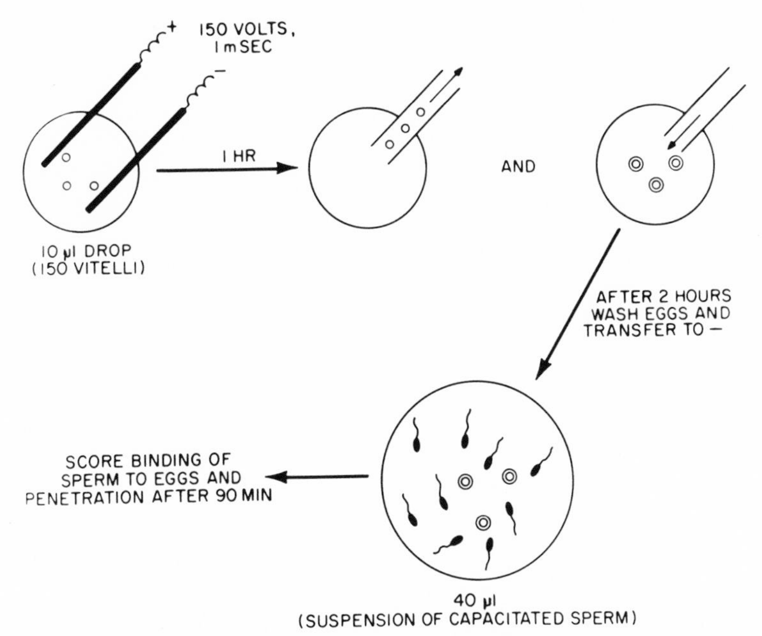

FIGURE 29. Outline of a procedure for collecting and assaying the cortical granule product of golden hamster eggs released by electrical stimulation.

mouse eggs, i.e., its action is not species specific, and it could be neutralized by inhibitors of pancreatic trypsin. These results suggest that the zona reaction is mediated, at least in part, by a trypsin-like serine protease, consistent with the sensitivity of the receptor-for-sperm in the zona pellucida to proteases demonstrated by Hartmann and Gwatkin (1971). Oikawa *et al*. (1975) have confirmed that pancreatic trypsin blocks the fertilization of hamster eggs *in vitro*, but they have also observed that eggs, exposed to trypsin for 25 min, fail to scatter light after exposure to wheat germ agglutinin (WGA), a change not seen in fertilized eggs obtained from the oviduct. However, since the cortical granule contents must diffuse across the perivitelline space and through the zona pellucida, insufficient active enzyme may reach the surface to produce all of the changes observed when the protease is applied externally. Oikawa *et al*. (1975), in fact, found that a loss of WGA binding did not occur when the time of exposure to trypsin was reduced to 10 min, a treatment nevertheless sufficient to block fertilization.

Further evidence that the zona reaction is produced by a protease from the cortical granules comes from the studies of Repin and Akimova (1976). These authors dissolved the zonae pellucidae of mice and rats with 1% sodium dodecylsulfate and subjected the solution to electrophoresis in capillaries filled with a 10–15% polyacrylamide gel. They found that after fertilization new protein bands appeared, indicating a partial hydrolysis of the zona proteins.

A trypsin-like protease, responsible for elevating the vitelline layer and inactivating the receptor-for-sperm, has also been demonstrated in the cortical granules of sea urchin eggs (Vacquier *et al*., 1972 a, b, 1973; Schuel *et al*., 1973; Troll *et al*., 1974; Longo *et al*., 1975). This enzyme appears to exist in the granules as a zymogen, since it is relatively inactive when liberated in the presence of a chelating agent (Vacquier, 1975a). The addition of Ca^{2+}, but not Mg^{2+}, causes a tenfold increase in enzyme activity. More recently, Carroll and Epel (1975) have resolved the protease into two components: one which breaks the attachments between the vitelline layer and the plasma membrane and the other which appears to alter the egg surface so that sperm no longer bind to it. These experiments were done by first precipitating the cortical granule exudate by adjusting the solution to pH 4.0, dissolving the precipitate in buffered saline, and applying the solution to a *p*-aminobenzamidine–Sepharose column. The protease activity was then eluted with 0.1 M NH_4OH–O.4 M NaCl. Early-eluting franctions of low specific activity (on α-*N*-benzoyl-L-arginine ethyl ester) delaminated the vitelline layer from the egg plasma membrane. Late-eluting fractions of high specific activity modified the vitelline layer surface, preventing sperm binding. The former they termed *vitelline delaminase* and the latter, *sperm receptor hydrolase*. The molecular weight of both enzymes was estimated to be 47,000 by sucrose gradient centrifugation. Carroll and Epel (1975) have suggested that the simultaneous release of these two proteases may explain the coincidence of membrane elevation and sperm detachment.

In addition to trypsin the cortical granules of sea urchin eggs have been shown to contain a β-1,3-glucanohydrolase (Epel *et al.*, 1969), which remains in solution on isoelectric precipitation of the proteases (Carroll and Epel, 1975). The role of this enzyme is unknown, but such glycosidases could be involved in dispersing the discharged cortical granule contents, which in both sea urchins and mammals are known to be rich in sulfated mucopolysaccharides (Schuel *et al.*, 1974). Dispersion of the cortical granule contents can be inhibited in sand dollar eggs by the addition of Concanavalin A (10 μg/ml) to the surrounding sea water (Vacquier and O'Dell, 1975). The effect of the lectin is blocked by α-D-methyl-mannopyranoside, indicating that it may be cross-linking terminal mannose residues in the cortical granule material.

In addition to the inactivation of receptor sites for sperm and the elevation of the vitelline layer which occur following fertilization in the sea urchin, there is a third change. This is a "hardening" of the vitelline layer that renders it insoluble in mercaptoethanol which dissolves the vitelline layer of unfertilized eggs (Lallier, 1970). This transition is prevented if the eggs are fertilized in the presence of penicillamine and semicarbazide, which appear to act by preventing the formation of cross-links, in addition to –S–S– bridges, between polypeptide chains, a result which suggests that hardening of the vitelline layer may be due to the formation of such new cross-links (Lallier, 1971). A counterpart to the hardening of the vitelline membrane appears to occur in the zona pellucida of mammals. Some authors have supposed that the increased resistance of the zona pellucida to digestion by proteases and mercaptoethanol is involved in the block to multiple sperm entry (Szollosi, 1967; Austin, 1961). However, this is clearly untrue, since hamster eggs exhibit a strong block to multiple sperm penetration of the zona pellucida, but no change in solubility properties, while rabbit eggs in which there is no such block exhibit a particularly marked resistance to solubilization by proteases and mercaptoethanol following fertilization (Table 6). It would appear that increased resistance to solubilization may be due to another agent released from the cortical granules which cross-links the peptide chains of zona proteins, possibly through the oxidation of lysine and other types of side chains (Lallier, 1971). This hardening or toughening of the zona pellucida is distinct from the block to multiple sperm entry, which involves the digestion of sperm receptor sites. It may be important in protecting the egg during cleavage as it is transported down the oviduct to the uterus. The presence of new intermolecular bonds would be expected to increase the resistance of the zona pellucida to protease digestion. Since the integrity of the zona pellucida following the cortical reaction would depend less heavily on disulfide bonding for its integrity, one would also expect increased insol-

TABLE 6. Block to Multiple Sperm Penetration of the Zona Pellucida and the Development of Resistance to Solubilization by Proteases and Mercaptoethanol

Species	Block to multiple sperm penetration of zona pellucida	Increased resistance of zona pellucida to solubilization in		
		Trypsin	Pronase	Mercaptoethanol
Mouse	+ (1, 2)	+ (3, 12)	+ (4, 12)	+ (5, 6)
Rat	+ (1, 2)	+ (7)		
Hamster	++ (1, 2)	− (7)	− (8)	− (8)
Rabbit	− (1)	++ (7, 9)	++ (10)	++ (9)
Sheep	++ (1)		− (11)	

(1) Braden *et al.* (1954); (2) Austin and Braden (1956); (3) Smithberg (1953); (4) Mintz and Gearhardt (1973); (5) Inoue and Wolf (1974); (6) Inoue and Wolf (1975); (7) Chang and Hunt (1956); (8) Gwatkin (unpublished); (9) Gould *et al.* (1971); (10) Fraser and Dandekar (1975); (11) Trounson and Moore (1974); (12) Krzanowska (1972).

ubility in mercaptans. The formation of such additional cross-linking may also be responsible for the slightly higher temperature required to melt the zonae pellucidae of mouse eggs (Chowlewa-Stewardt and Massaro, 1972) and their increased resistance to periodate digestion (Inoue and Wolf, 1974) following fertilization.

The studies of Wyrick *et al*. (1974) on the eggs of the South African clawed toad, *Xenopus laevis*, also point to the existence of a cross-linking agent. The cortical granule product from these eggs was shown to diffuse through the vitelline layer and, in the presence of Ca^{2+}, to form a precipitate with the innermost jelly layer (J1). This may be analogous to zona toughening. The cortical granule also contains a protease which may be responsible for blocking sperm binding (Hedrick, unpublished observations).

The block to multiple sperm penetration of the zona pellucida is dependent on the age of the eggs. Prior to ovulation the block appears to be relatively inefficient, perhaps because a suboptimal number of cortical granules have migrated to the periphery of the egg (Zamboni, 1970). Thus Iwamatsu and Chang (1971) found that mouse eggs collected from preovulatory follicles and fertilized *in vitro* exhibit a greater degree of polyspermy than recently ovulated eggs fertilized under the same conditions. Polyspermy also increases as eggs age after ovulation. Hamster eggs collected very soon after ovulation (17 hr post-HCG) were found to exhibit completely monospermic fertilization *in vitro* (Gwatkin, unpublished observations). However, when collected 21 and 24 hr after HCG means of 1.3 and 8.6 sperm per egg were obtained, respectively (Gwatkin and Williams, 1974). With further aging, hamster eggs discharge their cortical granule contents spontaneously and become resistant to sperm penetration (Yanagimachi and Chang, 1961). Spontaneous discharge does not occur in mouse, rat, and rabbit eggs, which become increasingly polyspermic with age (Szollosi, 1975).

Another important factor determining the efficiency of the block to polyspermy is the surrounding medium. Gwatkin and Williams (1974) found little or no polyspermy with recently ovulated hamster eggs fertilized in a modified Medium 199, supplemented with crystalline bovine serum albumin. However, Barros *et al*. (1972) reported that a high degree of polyspermy occurred when fertilization was carried out in a simple saline solution (0.15 M NaCl). Under such conditions either the cortical reaction does not occcur or the cortical granule exudate cannot act to inactivate the sperm receptor sites.

Chapter 12

Pronucleus Formation

Pronucleus formation has been studied under the electron microscope in the eggs of the golden hamster (Yanagimachi and Noda, 1970a), the mouse (Thompson *et al.*, 1974), and the pig (Szollosi and Hunter, 1973). This process has been reviewed recently by Longo (1973). The sperm nuclear membranes break down by the formation of multiple fusions between the inner and outer laminae, a process reminiscent of the vesiculation which occurs in the sperm acrosome reaction. Yanagimachi and Noda (1970a) have observed the sudden appearance of aggregates of small particles around the golden hamster sperm head at this time, the nature and function of which are not known. The chromatin, which consists of closely packed fibrils, then disperses and becomes separated from the remnant of the inner acrosomal membrane. The pronuclear envelope forms along the periphery of the dispersed chromatin as an aggregation of vesicles which later coalesce (Figure 30c,d). This process is similar to nuclear envelope formation in meiotic and mitotic cells (Chang and Gibley, 1968). Nucleoli develop, and the mitochondria appear to degenerate.

Meanwhile, the female chromosomes, formed when the second polar body is extruded at Anaphase II, disperse. Vesicles form around them and fuse to produce bilaminar envelopes,

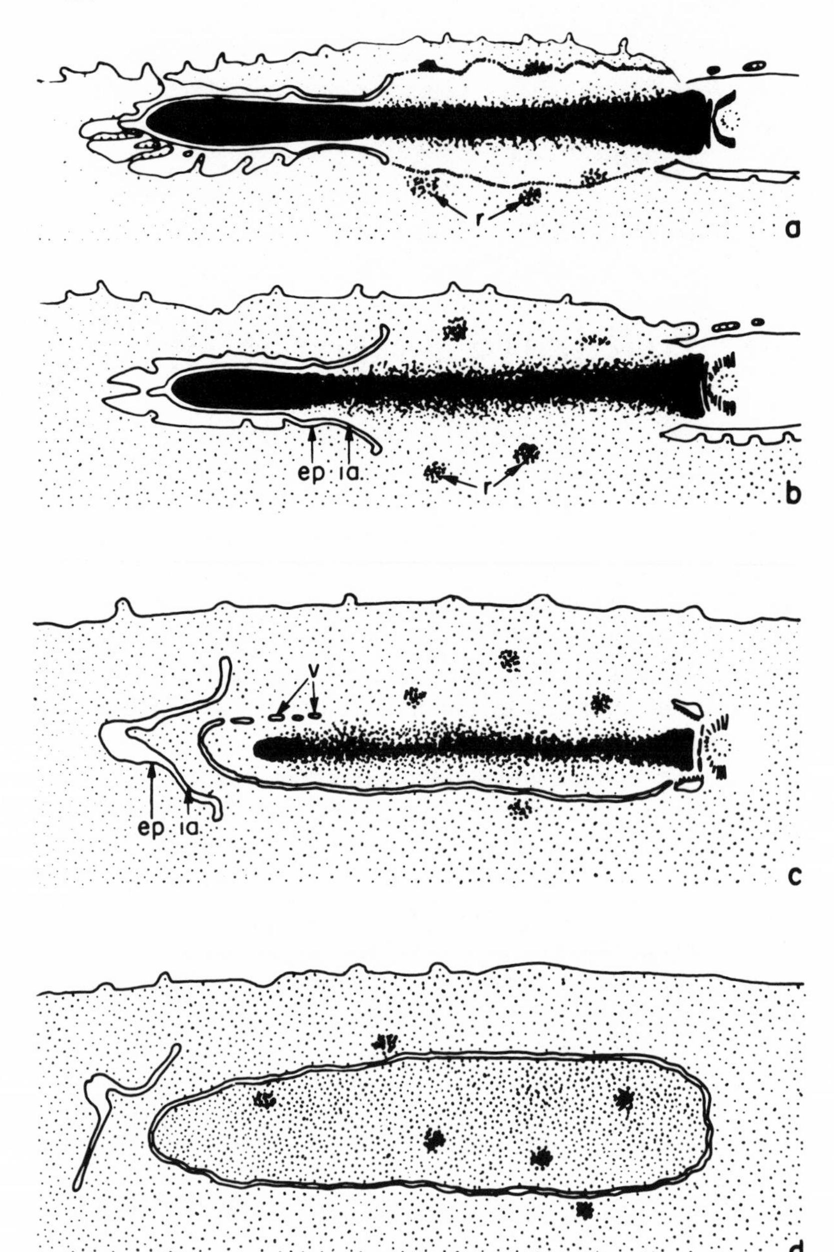
r
a
ep
ia
r
b
v
ep
ia
c
d

and these in turn fuse to form an irregular female pronucleus, which later becomes spherical and acquires nucleolus-like bodies. The male and female pronuclei eventually move together, and their bilaminar membranes interdigitate (Figure 31). Interdigitation, like microvilli, may facilitate fusion (Poste and Allison, 1973). The chromatin begins to condense and the pronuclear envelopes break down. The condensing chromosomes become associated with microtubules and intermix to form the metaphase of the first mitotic division. In mammals, which follow the *Ascaris*-type of fertilization (Longo, 1973), a zygote nuclear envelope, such as occurs in sea urchins, is not formed. The first time that the diploid chromosomes come to be surrounded by a nuclear envelope is at the two-cell stage.

During the formation of the male pronucleus, basic nuclear proteins characterized by a high content of arginine and cysteine are lost from the sperm (Kopecny and Pavlok, 1975). Autoradiographic studies on rabbit (Szollosi, 1966) and mouse (Luthardt and Donahue, 1973) eggs show that DNA synthesis by the sperm nucleus occurs more or less simultaneously with pronuclear formation.

Pronuclear formation can occur in hybrids, e.g., when capacitated mouse or rat sperm enter hamster or rabbit vitelli (Hanada and Chang, 1976). Intracellular conditions are critical. Thus, when rabbit sperm were fused to several types of somatic cells by Sendai virus the sperm nuclei failed to expand or to in-

←

FIGURE 30. Incorporation of the golden hamster sperm into the vitellus (redrawn from Yanagimachi and Noda, 1970a). (a) Breakdown of the nuclear membranes and the start of chromatin dispersion; (b) separation of the remnant of the inner acrosomal membrane from the dispersing chromatin; (c) formation of the pronuclear envelope; (d) chromatin dispersion and completion of the pronuclear envelope.

(ep) Plasma membrane of vitellus; (ia) inner acrosomal membrane of sperm; (r) aggregates of particles; (v) vesicles forming the pronuclear envelope.

a

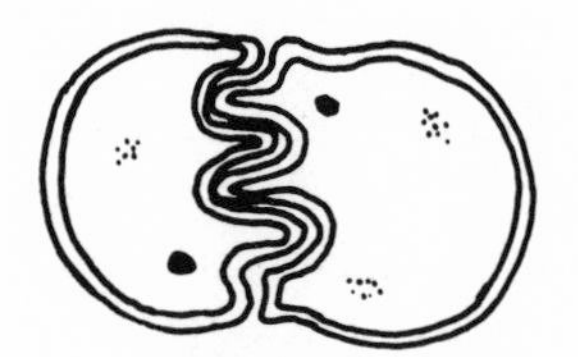

b

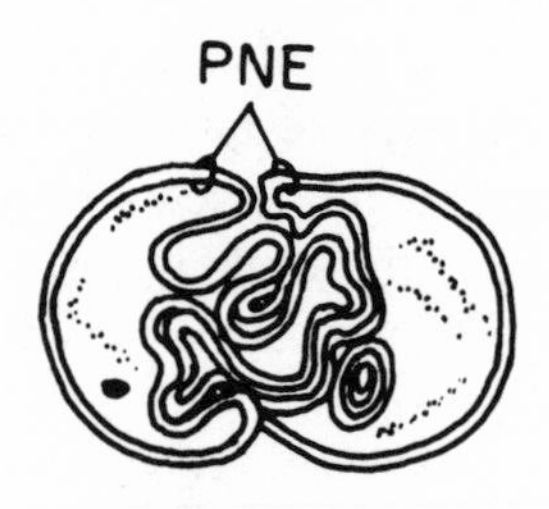

c

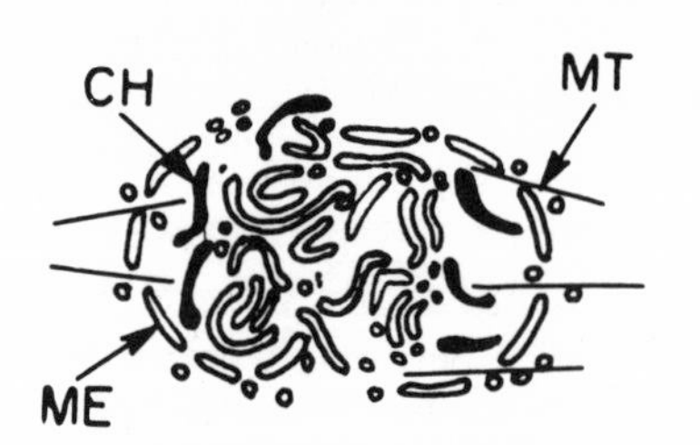

d

corporate [^{3}H] tymidine (Sawicki and Koprowski, 1971). The cytoplasmic conditions required to support sperm nucleus expansion and pronuclear formation are dependent on the stage of development of the oocyte. Sperm introduced into the sea urchin egg at the germinal vesicle stage do not form pronuclei (Franklin, 1965). The same is true for mouse (Iwamatsu and Chang, 1972) and hamster (Yanagimachi and Usui, 1972) eggs. As the germinal vesicle breaks down in the hamster the cytoplasm becomes capable of causing sperm nuclear decondensation (Yanagimachi and Usui, 1972). This property disappears soon after fertilization, reappears shortly before the first cleavage, and disappears again after its completion. Thibault (1973) and Thibault *et al*. (1975) have noted that rabbit and calf oocytes matured *in vitro* in the absence of added hormones are incapable of supporting pronuclear formation. However, when matured in the presence of gonadotropins, testosterone, and estradiol, or *in vivo*, they are capable of doing so. Taken together, these observations suggest that a factor is elaborated by the oocyte during normal maturation which controls pronucleus formation. Thibault (1973) has termed this hypothetical factor the *Male Pronucleus Growth Factor*, or MPGF. This factor could be a disulfide-reducing agent, since the highly condensed state of the sperm chromatin in eutherian mammals is due in part to disulfide cross-linkage of the nucleohistones (Bedford and Cal-

←

FIGURE 31. Association of rabbit male and female pronuclei (redrawn from Longo, 1973). (a) Interdigitation of pronuclear membranes; (b) infolding of pronuclear membranes and chromatin condensation; (c) vesiculation of pronuclear envelopes, formation of chromosomes and mitotic spindle; (d) first metaphase.

(CH) Chromosome; (ME) vesicle formed from fusion of inner and outer pronuclear membranes; (MT) microtubules of spindle; (PNE) pronuclear membranes.

vin, 1974; Wagner *et al.*, 1974). Mahi and Yanagimachi (1975) have suggested that MPGF could be reduced glutathione, which is abundant in cells and is known to fluctuate during the cell cycle.

Chapter 13

Metabolic Changes Associated with Fertilization

Because of the limited number of eggs available for experimentation, little is known of the metabolic changes which follow fertilization in mammals. However, considerable knowledge has been obtained with echinoderm eggs (Monroy, 1965, 1973; Epel, 1975).

In the sea urchin there is a reversal of membrane polarity 3–5 sec after insemination which is apparently due to an influx of Na^+ into the egg, since it is abolished by reducing the Na^+ concentration in the surrounding sea water to 10% of normal (Steinhardt *et al.*, 1971). Another change which may occur at about this time is an increase in calcium ions within the egg cytoplasm. Nakamura and Yasumasu (1974) have shown that dialyzable (i.e., free) calcium increases from O.1 mM in the unfertilized egg to 1 mM after fertilization. Steinhardt and Epel (1974) have found that eggs preloaded with ^{45}Ca show a twenty-fold increase in ^{45}Ca efflux when fertilized. This release also occurs when the eggs are treated with the ionophore A23187, which is known to transport divalent cations across the membranes of cells and subcellular organelles (Pressman, 1973; Schaffer *et al.*, 1974). This ionophore will also induce depolarization, the respiratory burst, the cortical reaction, and all of the later metabolic changes which occur approximately 5 min after insemination (see Figure 32). Treatment of hamster eggs with A23187 will, in addition, induce parthenogenetic development, at least to the pronuclear stage (Steinhardt *et al.*, 1974). Clearly the rise in intracellular calcium ions is a key factor in metabolic activation.

The release of intracellular calcium could be responsible for depolarization of the egg membrane. Binding of Ca^{2+} to one side of a phosphatidylserine bilayer has been observed to alter the polarity of this synthetic membrane (Wobschall and Ohki, 1973). Calcium ions are known to activate a number of enzymes, e.g., phosphorylase (Ozawa *et al.*, 1967), involved in

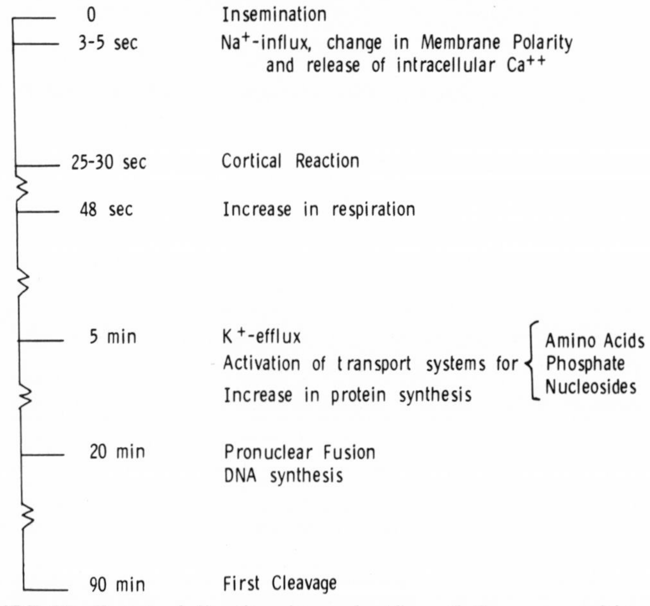

FIGURE 32. Events following insemination of the sea urchin egg. After Epel, 1975.

respiration, and it may be that this activation is responsible for the burst in respiration that occurs 48 sec after the insemination of sea urchin eggs.

Approximately 4 min after the burst in respiration several other processes begin. These include plasma membrane changes which result in the development of a K^+ efflux (Steinhardt *et al.*, 1971); the transport of amino acids (Epel, 1972), phosphate (Whiteley and Chambers, 1966), and nucleosides (Piatigorsky and Whiteley, 1965); and a large increase in protein synthesis (Epel, 1967). Some of the so-called "late" changes (K^+ efflux, protein synthesis, and DNA synthesis) can be induced in the absence of sperm by 1 mM ammonia without the occurrence of the early changes (Na^+ influx, the cortical reaction, and the respiratory burst), showing that the late changes are programmed independently of the early ones (Epel *et al.*, 1974). Neither are all of the late changes causally related, since Epel *et al.* showed that ammonia failed to activate amino acid transport, and when the K^+ efflux was abolished by acidifying the sea water the increase in protein synthesis still occurred.

The development of a K^+ conductance and the development of transport systems may involve the release of one or more peripheral proteins from the oolemma which regulate these processes. Thus, when the surface proteins of unfertilized sea urchin eggs were labeled with ^{125}I by a lactoperoxidase procedure and the eggs were fertilized or activated by ammonia or A23187, 15–25% of the total labeled protein was released, most of it as a 150,000-dalton glycoprotein (Johnson and Epel, 1975). When the released material was dialyzed, concentrated, and added back to partially activated eggs, the rate of protein synthesis was suppressed to the level of the unactivated eggs.

In the sea urchin pronuclear fusion occurs 20 min after insemination and is followed rapidly by DNA synthesis. In the eggs of primitive chordates, e.g., the ascidians, DNA synthesis is accompanied by a sharp rise in the number of Concanavalin A binding sites on the egg surface (Monroy *et al.*, 1973), suggesting that the stimulus for DNA replication may originate in the plasma membrane. This in turn may be mediated by a cytoplasmic factor, probably DNA polymerase, as has been demonstrated by nuclear transplant experiments with amphibian eggs (Gurdon and Woodland, 1968).

Chapter 14

Fate of Nonfertilizing Spermatozoa and Interaction of Spermatozoa with Somatic Cells

Of the millions of mammalian sperm inseminated into the female genital tract, only a very few ever reach the site of fertilization. The major mechanism by which excess spermatozoa are eliminated is by leukocytes that invade the uterus (Austin, 1960; Yanagimachi and Chang, 1963b; Bedford, 1965; Mahajan and Menge, 1966; Moyer *et al.*, 1967) and uterine cervix (Moyer *et al.*, 1970) after copulation. Phagocytosis of spermatozoa by epithelial cells of the endometrium (Moyer *et al.*, 1967) and oviducal mucosa (Austin, 1960; Zamboni, 1971a) has also been reported. Nonfertilizing sperm persisting in the female tract may also occasionally be phagocytized by embryos at the two-cell (Thompson and Zamboni, 1974) and blastocyst (Tachi and Kraicer, 1967; McReynolds and Hadek, 1971) stages. In rodents and lagomorphs clearance is rapid, occurring within 24–36 hr after copulation (Howe, 1967; Bishop, 1969). However, little is known concerning the clearance from the female tract of spermatozoa in species such as the dog, horse, and ferret in which sperm survive much longer, a property possibly correlated with the longer estrus period in these mammals.

Coppelson and Reid (1967, 1974) have proposed that uptake of nonfertilizing spermatozoa by epithelial cells in the cervix and uterus could result in transfer of genetic information from the sperm to these cells with possible risk of neoplastic transformation.

Several studies have been made over the past few years on the interaction of mammalian sperm with mammalian somatic cells cultured *in vitro*. Despite an initial negative report (Sawicki and Koprowski, 1971), evidence has been obtained to show that sperm nuclei can undergo limited activation, including DNA synthesis, within these somatic cells (Croce *et al.*, 1972; Zelenin *et al.*, 1974). Treatment of Chinese hamster DON cells with rat sperm has been claimed to result in the subsequent expression of rat fetal antigens by these cells (Higgins *et al.*, 1975). However, karyotypic analysis failed to reveal the presence of rat chromosomes, indicating that an extremely limited amount of genetic material had been transferred. Nevertheless, even such limited transfer is of great interest, and if the technique could be refined it might ultimately permit the direct mapping of genes in a haploid genome.

Chapter 15

Parthenogenesis

The unfertilized eggs of many mammals have a tendency to develop spontaneously in the absence of a male gamete (Beatty, 1967; Tarkowski, 1971; Graham, 1974). This phenomenon is termed *parthenogenesis* and is of considerable significance for a full understanding of the role of the sperm in fertilization. If the second polar body is retained the parthenote is diploid. Haploid parthenotes are formed if the second polar body is extruded, or if the egg cleaves immediately so that the pronucleus is in one cell and the second polar body nucleus is in the other (Figure 33). Spontaneous parthenogenesis in mammals does not usually proceed to mitosis.

Both the proportion of eggs activated and the stage of development achieved can be enhanced by temperature shock (Pincus and Enzmann, 1936; Pincus and Shapiro, 1940; Thibault, 1947; Chang, 1954; Austin, 1956b; Komar, 1973), osmotic shock (Pincus and Enzmann, 1936; Pincus, 1939), removal of the cumulus oophorus with hyaluronidase (Graham, 1970), electrical stimulation (Tarkowski *et al*., 1970), and exposure to the divalent ionophore, A23187 (Steinhardt *et al*., 1974).

The most extensive development of a mammalian parthenote has been obtained after electrical stimulation of mouse eggs

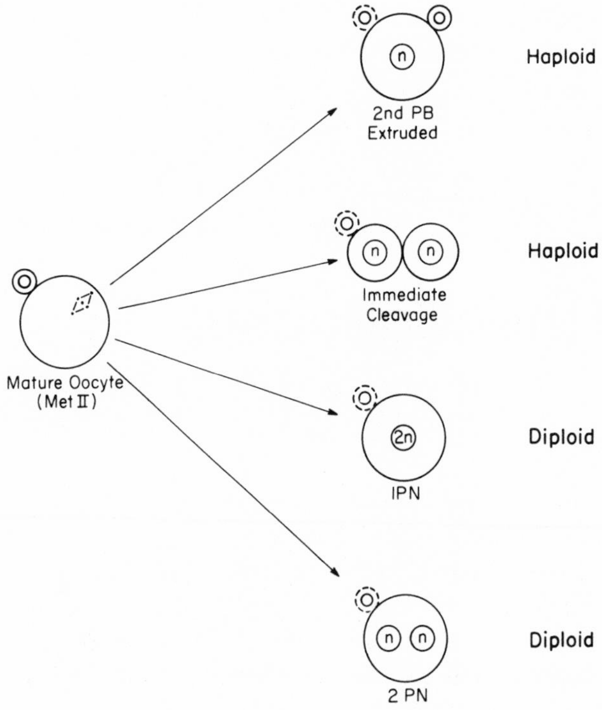

FIGURE 33. The principal types of mammalian parthenogenetic development.

within the oviduct. Forty-six percent of the eggs were activated, the most common reaction being extrusion of the second polar body and the formation of one pronuclueus (Witkowska, 1973a,b). Seventy percent of the activated eggs were haploid. The number of developing embryos declined each day, so that by the tenth day of gestation only 2.3% remained. However, a few developed to the egg-cylinder (Figure 34) and eight-somite stages.

By removing the cumulus oophorus from mouse eggs with hyaluronidase, Graham (1970) obtained a number of parthenotes which developed into blastocysts *in vitro*. After transplantation beneath the kidney capsule, growths were obtained which yielded a haploid cell strain on cell culture. Kaufman and Surani

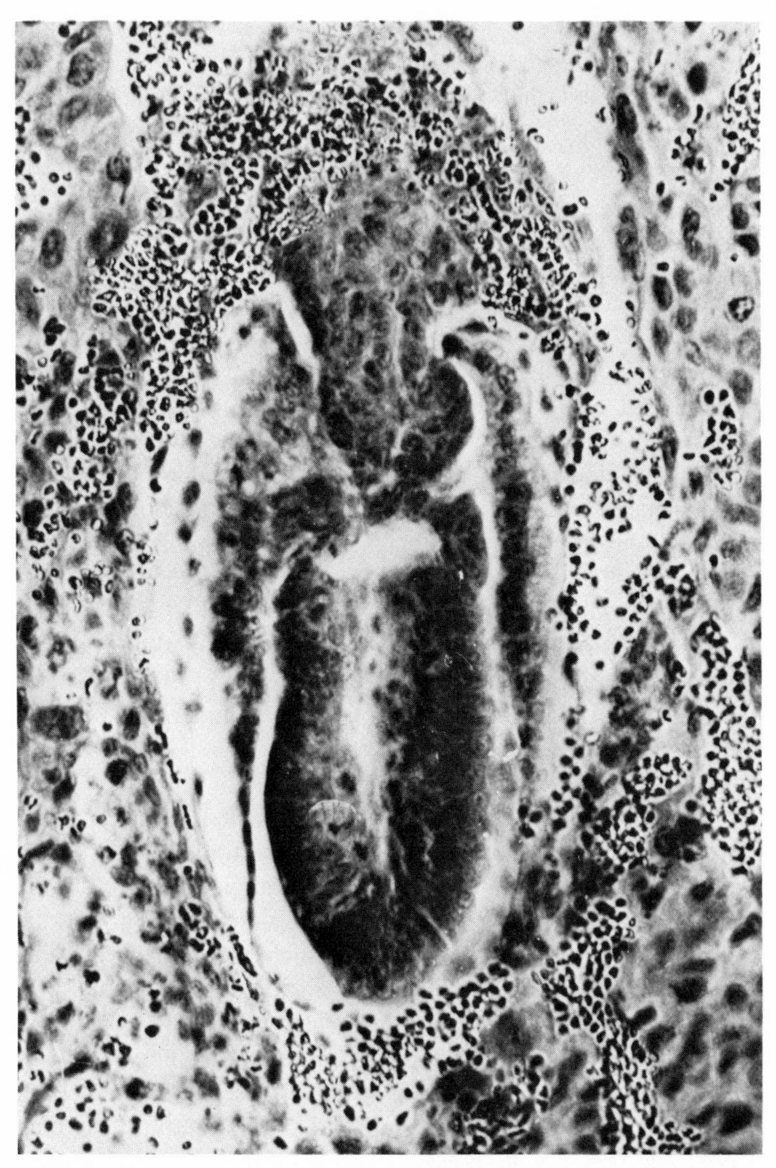

FIGURE 34. Section through a parthenogenetic mouse embryo which developed in the uterus after electrical stimulation of eggs in the oviduct. This egg-cylinder stage corresponds to approximately 7 days of pregnancy. X400 (reduced 10% for reproduction). (Tarkowski *et al.*, 1970.)

(1974) observed that the cultivation of hyaluronidase-treated mouse eggs in a medium of low osmolarity inhibited extrusion of the second polar body so that the proportion of haploid parthenotes was reduced.

A 2-min exposure of hamster eggs to 3 μM A23187 produces cortical granule discharge and induces 70% of the eggs to form pronuclei (Steinhardt *et al.*, 1974). A23187 also activates sea urchin eggs, producing a cortical reaction, plasma membrane conductance changes, a respiratory burst, and an increase in protein and DNA synthesis (Steinhardt and Epel, 1974). The compound appears to act by releasing intracellular stores of Ca^{2+} rather than by increasing the transport of Ca^{2+} across the plasma membrane, since its effect is independent of the ionic composition of the surrounding medium. Golden hamster eggs undergo parthenogenetic cleavage *in vitro* after electrical stimulation (Gwatkin, unpublished observation).

It seems likely that all of these activation methods may be altering the egg plasma membrane. Pienkowski (1974) has demonstrated that the vitelli of mouse eggs undergo a surface change after hyaluronidase activation, so that they are agglutinated at the same low concentration of Concanavalin A (10 μg/ml) as the vitelli of fertilized eggs. The vitelli of unfertilized eggs, in contrast, require a 2 mg/ml concentration of the lectin for agglutination. This membrane modification could lead to the loss of a cytostatic factor in the egg cytoplasm which arrests the chromosomes in Metaphase II. No such factor has yet been demonstrated in mammals, but Masui and Markert (1971) have reported that the injection of cytoplasm from frog oocytes in Metaphase II into the blastomeres of two-cell embryos arrests them in metaphase. Cytoplasm from fertilized eggs is ineffective. The factor is a macromolecule, apparently containing RNA, and requires the presence of Ca^{2+} for its activity (Masui, 1974).

It is also likely that a cortical reaction is required for advanced parthenogenetic development. A cortical reaction occurs after electrical stimulation (Gwatkin *et al.*, 1973b), but not after cold shock (Gulyas, 1974; Longo, 1975), hyaluronidase treatment, or osmotic shock (Solter *et al.*, 1974) and may explain the relatively extensive development obtained after electrical stimulation (Tarkowski *et al.*, 1970). Mintz and Gearhardt (1973) have suggested that even electrical stimulation may not be an entirely adequate substitute for sperm penetration, since the zonae pellucidae of mouse parthenotes obtained by this means exhibit a digestion time in pronase which was intermediate between that of unfertilized and fertilized eggs.

The reason for the death of parthenotes, even after an apparently normal cortical reaction, is unknown. It has been suggested that their demise could be due to the unmasking of recessive lethal genes. However, Graham (1974) has pointed out that inbreeding of mouse strains, which has been assumed to eliminate such genes, does not prevent the early death of parthenotes developed from them. Nor do fused haploid blastomers from different mouse strains fare any better (Graham, 1971). Failure of parthenotes to develop in the uterus cannot be due to their inability to cytodifferentiate, since when haploid mouse blastocysts, obtained by hyaluronidase activation, were transferred beneath the testis capsule they differentiated into a wide range of cell types, including neural, cartilage, bone, and muscle cells (Iles *et al.*, 1975). Possibly normal development *in utero* may depend on a specific product of the male genome.

In the future it is anticipated that improved techniques for producing parthenotes will be developed. These could then be used to assess the role of the spermatozoon in development, to determine whether it may act as a vector to carry viruses or virus-like particles into the egg (Calarco and Szollosi, 1973; Chase and Pikó, 1973), and in genetic studies, as a source of haploid cells in which all mutations would be expressed. Research in parthenogenesis should be facilitated by the recent discovery of the LT mouse strain, in which half of the females exhibit spontaneous parthenogenetic cleavage of eggs within the ovary at 2–4 months of age, some forming teratomas (Stevens and Varnum, 1974).

Epilogue

Perhaps the best way to conclude this review of recent advances in our knowledge of mammalian fertilization is to outline what we still need to learn. This has been done as a series of questions, many of which it is hoped will be answered in the next decade.

1. How is oocyte maturation controlled? Is the oocyte maturation inhibitor (OMI) present in follicular fluid responsible? Does the concentration of this inhibitor decline, or does LH act directly on the oocyte membrane to overcome the inhibition? Is maturation mediated by a specific cytoplasmic factor? What changes occur in the membrane of the egg as it ages in the oviduct?
2. What specific changes occur in the structure of various regions of the sperm plasma membrane during maturation in the epididymis? Do they result from absorption, insertion, unmasking, or chemical alteration of existing membrane components?
3. What are the specific factors regulating gamete transport to the site of fertilization? What component of the cumulus oophorus induces oviduct contractions and appears to control the number of sperm moving from the isthmus to the site of fertilization in the ampulla?

4. What are the specific ionic and nutritional requirements for successful fertilization of various mammalian species *in vitro*? Does albumin stabilize the gamete membranes, chelate toxic ions, or have some other function? Can improved media be developed to support fertilization in a wider range of species with the same efficiency and reliability as *in vivo*? Are such fertilizations normal, as would be established by the birth of healthy young on transplantation of the embryos to foster mothers?
5. What specific seminal plasma and membrane glycoproteins must be removed from the sperm to induce capacitation? Can ultrastructural techniques of sufficient resolution be developed to reveal specific changes in the membrane?
6. Can a method be developed for routinely capacitating large numbers of sperm for biochemical studies? Would a mixture of glycosidases suffice?
7. Is cumulus cell action sufficient to account for capacitation *in vivo*, or are the oviduct and uterus also required? Are there distinct species variations in the site of capacitation?
8. Is capacitation essentially an enzymatic process, or are other processes also involved, e.g., hypertonic conditions?
9. What is the nature of the motility factor supplied to the sperm by the cumulus oophorus?
10. How is the change in sperm motility from a progressive one to a bobbing motion of the head brought about?
11. Does the sperm bind to the zona pellucida by a specific region of its plasma membrane? Is this limited to the region overlying the acrosome?
12. What is the nature of the factor from the vitellus which appears to control sperm binding to the egg?
13. What is the nature of the receptor-for-sperm in the zona pellucida? Is it a glycoprotein and is the sugar moiety critical? What is the active site? What is the nature of the complementary receptor on the sperm?
14. Are metabolic changes in the sperm needed for the acrosome reaction to take place? Does acrosin play a role? How does Ca^{2+} act to modify the sperm membrane?

15. Is acrosin located over the whole inner acrosomal membrane or only in the equatorial region? Is it responsible for penetration of the zona pellucida or is another acrosomal enzyme involved?
16. What is the reason for the inability of the inner acrosomal membrane to fuse with the oolemma?
17. How is the fertilization cone formed? Are microfilaments involved?
18. What specific change occurs in the oocyte membrane following maturation which promotes gamete fusion?
19. What is the nature of the serine protease released from cortical granules? Can it be assayed on synthetic substrates?
20. What specific rearrangements occur in the oolemma following fertilization to prevent subsequent sperm fusions?
21. How is the receptor-for-sperm in the zona pellucida altered by the cortical granule exudate to prevent subsequent sperm binding?
22. What type of cross-linking occurs in the zona pellucida of some species following fertilization? How important is the change for embryo transport and implantation?
23. What specific factor in the egg is responsible for sperm pronucleus formation? Is it a desulfide-reducing agent?
24. What metabolic events are activated in mammalian eggs by fertilization? Do they differ from those in echinoderms? Does the release of Ca^{2+} induce these changes, and are they linked or independent? Does a peripheral glycoprotein suppress metabolism in the unfertilized mammalian egg, and is its action mediated by a cytoplasmic factor?
25. What is the cause of the death of parthenotes shortly after implantation? Does the sperm supply a factor necessary for development in the uterus?

The knowledge gained in attempting to answer these questions should lead to a number of applications in medicine and biology. *In vitro* fertilization techniques might be used to diagnose cases of human infertility involving abnormal gamete physiology (Overstreet, 1975) or to assess the fertility of the semen used to artificially inseminate livestock. Another important application lies in the development of new nonhormonal contraceptives which interfere with the mechanisms involved in the fertilization process (Gwatkin, 1975).

In addition to these practical applications, fertilization presents the experimental biologist with several basic biological processes, which can be studied *in vitro* under relatively defined conditions. The binding reactions between the uncapacitated spermatozoon and the egg provides a model for understanding the mechanisms involved in cell recognition and how they may operate in differentiation and immunity, processes considerably less accessible to experimentation. The cortical reaction provides a model for studying membrane fusion. Understanding the means by which the sperm, or a parthenogenetic stimulus, overcomes the arrest of the egg in Metaphase II may lead to an understanding of the mechanisms which control cell growth and malignancy. Although relatively small numbers of eggs can be obtained from mammals so that new microtechniques will be required, the data provided by studies with mammalian fertilization systems are probably more directly applicable to human and veterinary medicine than thos obtained from studies with invertebrate gametes. A further consideration is that mammalian fertilization processes occur much more slowly than those in lower forms and hence can be more easily analyzed.

References

Adams, C. E. (1956). A study of fertilization in the rabbit: The effect of post-coital ligation of the fallopian tube or uterine horn. *J. Endocrinol.* **13,** 296–308.

Aketa, K. (1967). On sperm-egg bonding as the initial step of fertilization in the sea urchin. *Embryologia* **9,** 238–245.

Aketa, K. (1973). Physiological studies on the egg surface component responsible for sperm–egg bonding in sea urchin fertilization. I. Effect of sperm-binding protein on the fertilizing capacity of sperm. *Exp. Cell Res.* **80,** 439–441.

Aketa, K. (1975). Physiological studies on the sperm surface component responsible for sperm–egg bonding in sea urchin fertilization. II. Effect of Concanavalin A on the capacity of sperm. *Exp. Cell Res.* **90,** 56–62.

Aketa, K., and Tsuzuki, H. (1968). Sperm-binding capacity of the S—S reduced protein of the vitelline membrane of the sea urchin egg. *Exp. Cell Res.* **50,** 675–676.

Aketa, K., Tsuzuki, H., and Onitake, K. (1968). Characterization of the sperm-binding protein from sea urchin egg surface. *Exp. Cell Res.* **50,** 676–679.

Aketa, K., Onitake, K., and Tsuzuki, H. (1972). Tryptic disruption of sperm-binding site of sea urchin egg surface. *Exp. Cell Res.* **71,** 27–32.

Allison, A. C., and Hartree, E. F. (1970). Lysosomal enzymes in the acrosome and their possible role in fertilization. *J. Reprod. Fertil.* **21,** 501–515.

Andersen, E. (1974). Comparative aspects of the ultrastructure of the female gamete. *Int. Rev. Cytol., Suppl.* **4,** 1–70.

Andersen, E., Hoppe, P. C., Whitten, W. K., and Lee, G. S. (1975). *In vitro* fertilization and early embryogenesis: A cytological analysis. *J. Ultrastruct. Res.* **50,** 231–252.

Austin, C. R. (1948). Number of sperms required for fertilization. *Nature (London)* **162,** 534–535.

Austin, C. R. (1951). Observations on the penetration of sperm into the mammalian egg. *Austral. J. Sci. Res.* (B) **4,** 581–592.

Austin, C. R. (1952). The capacitation of mammalian sperm. *Nature (London)* **170,** 326.

Austin, C. R. (1953). The growth of knowledge on mammalian fertilization. *Austral. Vet. J.* **29,** 191–198.

Austin, C. R. (1956a). Cortical granules in hamster eggs. *Exp. Cell Res.* **10,** 533–540.

Austin, C. R. (1956b). Activation of eggs by hypothermia in rats and hamsters. *J. Exp. Biol.* **33,** 338–347.

Austin, C. R. (1960). Fate of spermatozoa in the female genital tract. *J. Reprod. Fertil.* **1,** 151–156.

Austin, C. R. (1961). *The Mammalian Egg.* Charles C Thomas, Springfield, Ill.

Austin, C. R. (1965). Fertilization. *Foundations of Developmental Biology Series.* Prentice-Hall, Englewood Cliffs, N. J.

Austin, C. R. (1968). *Ultrastructure of Fertilization.* Holt, Rhinehart & Winston, New York.

Austin, C. R. (1974). Principles of fertilization. *Proc. Royal Soc. Med.* **67,** 925–927.

Austin, C. R., and Bishop, M.W.H. (1958). Some features of the acrosome and perforatorium in mammalian spermatozoa. *Proc. Royal Soc. Series B.* **149,** 234–240.

Austin, C. R., and Braden, A.W.H. (1956). Early reactions of the rodent egg to spermatozoon penetration. *J. Exp. Biol.* **33,** 358–365.

Austin, C. R., and Lovelock, J. E. (1958). Permeability of rabbit, rat and hamster egg membranes. *Exp. Cell Res.* **15,** 260–261.

Austin, C. R., and Short, R. V., eds. (1972). *Germ Cells and Fertilization,* Book 1, *Reproduction in Mammals.* Cambridge Univ. Press, Cambridge.

Austin, C. R., and Smiles, J. (1948). Phase-contrast microscopy in the study of fertilization and early development of the rat egg. *J. Royal Micr. Soc.* **68,** 13–19.

Ayalon, D., Tsafriri, F., Lindner, H. R., Cordova, T., and Harell, A. (1972). Serum gonadotrophin levels in proestrus rats in relation to the resumption of meiosis by the oocytes. *J. Reprod. Fertil.* **31,** 51–58.

Bae, I., and Foote, R. H. (1975). Carbohydrate and amino acid requirements and ammonia production of rabbit follicular oocytes matured *in vitro*. *Exp. Cell Res.* **91,** 113–118.

Baker, T. G. (1972). Oogenesis and ovarian development. In *Reproductive Biology* (H. Balin and S. Glasser, eds.) pp. 398–437. Excerpta Medica, Amsterdam.

Baker, T. G., and Franchi, L. L. (1967). The structure of the chromosomes in human primordial oocytes. *Chromosoma (Berlin)* **22,** 358–377.

Baker, T. G., and Neal, T. G. (1970). Gonadotrophin-induced maturation of mouse Graffian follicles in organ culture. In *Ooogenesis* (J. D. Biggers and A. W. Schuetz, eds.), pp. 377–396. University Park Press, Baltimore.

Barlow, P., and Vosa, C. G. (1970). The Y chromosome in human spermatozoa. *Nature (London)* **226,** 961–962.

Barros, C. (1967). *In vitro* fertilization and the sperm acrosome reaction in the hamster. *J. Exp. Zool.* **166,** 317–324.

Barros, C. (1968). *In vitro* capacitation of golden hamster spermatozoa with fallopian tube fluid of the mouse and rat. *J. Reprod. Fertil.* **17,** 203–206.

Barros, C. (1974). Capacitation of mammalian spermatozoa. In *Physiology and Genetics of Reproduction* (E. M. Coutinho and F. Fuchs, eds.), Part B, pp. 3–24. Plenum Press, New York.

Barros, C., and Garavagno, A. (1970). Capacitation of hamster spermatozoa with blood sera. *J. Reprod. Fert.* **22,** 381–384.

Barros, C., and Munoz, G. (1973). Sperm–egg interaction in immature hamster oocytes. *J. Exp. Zool.* **186,** 73–78.

Barros, C., and Yanagimachi, R. (1971). Induction of zona reaction in golden hamster eggs by cortical granule material. *Nature (London)* **233,** 268–269.

Barros, C., and Yanagimachi, R. (1972). Polyspermy-preventing mechanisms in the golden hamster egg. *J. Exp. Zool.* **180,** 251–266.

Barros, C., Bedford, J. M., Franklin, L. E., and Austin, C. R. (1967). Membrane vesiculation as a feature of the mammalian acrosome reaction. *J. Cell Biol.* **34,** C1–5.

Barros, C., Vliegenthart, A. M., and Franklin, L. E. (1972). Polyspermic fertilization of hamster eggs *in vitro*. *J. Reprod. Fertil.* **28,** 117–120.

Barros, C., Fujimoto, M., and Yanagimachi, R. (1973). Failure of zona penetration of hamster spermatozoa after prolonged preincubation in a blood serum fraction. *J. Reprod. Fertil.* **35,** 89–95.

Barry, M. (1843). Spermatozoa observed within the mammiferous ovum. *Phil. Trans. Royal Soc., London* **133,** 33.

Bavister, B. D., and Morton, D. B. (1974). Separation of human serum components capable of inducing the acrosome reaction in hamster spermatozoa. *J. Reprod. Fertil.* **40,** 495–498.

Bavister, B. D., Yanagimachi, R., and Teichman, R. J. (1976). Capacitation of hamster spermatozoa with adrenal gland extracts. *Biol. Reprod.* **14,** 219–221.

Beatty, R. A. (1967). Parthenogenesis in vertebrates. In *Fertilization* (C. Metz and A. Monroy, eds.) Vol. 1, Chap. 9, pp. 413–440. Academic Press, New York.

Bedford, J. M. (1963). Changes in the electrophoretic properties of rabbit spermatozoa during passage through the epididymis. *Nature* (*London*) **200,** 1178–1180.

Bedford, J. M. (1965). Effect of environment on phagocytosis of rabbit spermatozoa. *J. Reprod. Fertil.* **9,** 249–256.

Bedford, J. M. (1967). Observations on the fine structure of spermatozoa of the bush baby (*Galago senegalensis*), the African green monkey (*Cercopithecus aethiops*) and man. *Amer. J. Anat.* **121,** 443–459.

Bedford, J. M. (1968). Ultrastructural changes in the sperm head during fertilization in the rabbit. *Amer. J. Anat.* **123,** 329–358.

Bedford, J. M. (1969). Morphological aspects of sperm capacitation in mammals. *Adv. Biosci.* **4,** 35–50.

Bedford, J. M. (1970). Sperm capacitation and fertilization in mammals. *Biol. Reprod., Suppl.* **2,** 128–158.

Beford, J. M. (1972). Sperm transport, capacitation and fertilization. In *Reproductive Biology* (H. Balin and S. Glasser, eds.), pp. 338–392, Excerpta Medica, Amsterdam.

Bedford, J. M., and Calvin, H. I. (1974). The occurrence and possible functional significance of —S—S— crosslinks in sperm heads, with particular reference to eutherian mammals. *J. Exp. Zool.* **188,** 137–156.

Bedford, J. M., and Chang, M. C. (1962). Removal of decapacitation factor from seminal plasma by high-speed centrifugation. *Amer. J. Physiol.* **202,** 179–181.

Bedford, J. M., Cooper, G. W., and Calvin, H. I. (1973). Postmeiotic changes in the nucleus and membranes of mammalian spermatozoa. In *Genetics of the Spermatozoon* (R. A. Beatty and S. Glueckson-Waelsch, eds.), pp. 69–89. Univ. of Edinburgh, and Einstein College of Medicine, New York.

Berlin, R. D., and Ukena, T. E. (1972). Effect of colchicine and vinblastine on the agglutination of polymorphonuclear leucocytes by Concanavalin A. *Nature* (*London*) **238,** 120–122.

Bhargava, P. M., Bishop, M.W.H., and Work, T. S. (1959). The chemical composition of bull semen with special reference to nucleic acids, free nucleotides and free amino acids. *Biochem. J.* **73,** 242–247.

Bishop, D. W. (1969). Sperm physiology in relation to the oviduct. In *The Mammalian Oviduct* (E. S. E. Hafez and R. J. Blandau, eds.), pp. 231–250. Univ. of Chicago Press, Chicago.

Bishop, M. W. H., and Walton, A. (1960). Spermatogenesis and the structure of mammalian spermatozoa. In *Marshall's Physiology of Reproduction* (A. S. Parkes, ed.), Vol. 1, Chap. 7. Longmans Green & Co., London.

Blandau, R. J. (1969). Gamete transport-comparative aspects, In *The Mammalian Oviduct* (E. S. E. Hafez, and R. J. Blandau, eds.), pp. 129–162. Univ. of Chicago Press, Chicago.

Blandau, R. J., and Jordan, E. S. (1941). The effect of delayed fertilization on the development of the rat ovum. *Amer. J. Anat.* **68,** 275–291.

Blandau, R. J., and Money, W. L. (1944). Observations on the rate of transport of spermatozoa in the female genital tract of the rat. *Anat. Rec.* **90,** 255–279.

Blandau, R. J., and Odor, D. L. (1972). Observations on the behavior of oogonia and oocytes in tissue and organ culture. In *Oogenesis* (J. D. Biggers and A. W. Schuetz, eds.), pp. 301–320. University Park Press, Baltimore.

Blandau, R. J., and Young, W. C. (1939). The effects of delayed fertilization on the development of the guinea pig ovum. *Amer. J. Anat.* **64,** 303–329.

Bleau, G., Bodley, F., Longpré, J., Capdelaine, A., and Roberts, K. D. (1974). Cholesterol sulfate. I. Occurrence and possible biological function as an amphipathic lipid in the membrane of the human erythrocyte. *Biochim. Biophys. Acta* **352,** 1–8.

Bleau, G., VandenHeuvel, W. J. A., Andersen, O. F., and Gwatkin, R. B. L. (1975). Desmosteryl sulphate of hamster spermatozoa, a potent inhibitor of capacitation *in vitro. J. Reprod. Fertil.* **43,** 175–178.

Borrell, U., Nilsson, O., and Westman, A. (1957). Ciliary activity in the rabbit fallopian tubes during oestrus and after copulation. *Acta Obstet. Gynecol. Scand.* **36,** 22–28.

Brackett, B. G. (1969). Effect of washing the gametes on fertilization *in vitro. Fertil. Steril.* **20,** 127–141.

Brackett, B. G., and Oliphant, G. (1975). Capacitation of rabbit spermatozoa *in vitro. Biol. Reprod.* **12,** 260–274.

Brackett, B. G., and Williams, W. L. (1968). Fertilization of rabbit ova in a defined medium. *Fertil. Steril.* **19,** 144–155.

Braden, A. W. H. (1952). Properties of the membranes of rat and rabbit eggs. *Austral. J. Sci. Res.* (B) **5,** 460–471.

Braden, A. W. H. (1953). Distribution of sperms in the genital tract of the female rabbit after coitus. *Austral. J. Biol. Sci.* **6,** 693–705.

Braden, A. W. H. and Austin, C. R. (1954). The number of sperms about the eggs in mammals and its significance for normal fertilization. *Austral. J. Biol. Sci.* **7,** 543–551.

Braden, A. W. H., Austin, C. R., and David, H. A. (1954). The reaction of the zona pellucida to sperm penetration. *Austral. J. Biol. Sci.* **7,** 391–409.

Briggs, M. H. (1973). Steroid hormones and the fertilizing capacity of spermatozoa. *Steroids* **22,** 547–553.

Brown, C. R. (1975). Distribution of hyaluronidase in the ram spermatozoon. *J. Reprod. Fertil.* **45,** 537–539.

Brown, C. R., and Hartree, E. F. (1974). Distribution of a trypsin-like proteinase in the ram spermatozoon. *J. Reprod. Fertil.* **39,** 195–198.

Brown, C. R., and Hartree, E. F. (1976). Effects of acrosin inhibitors on the soluble and membrane-bound forms of ram acrosin, and a reappraisal of the role of the enzyme in fertilization. *Z. Physiol. Chem.* **357,** 57–65.

Brown, C. R., Andani, Z., and Hartree, E. F. (1975). Studies on ram acrosin: Isolation from spermatozoa, activation by cations and organic solvents and influence of cations on its reaction with inhibitors. *Biochem. J.* **149,** 133–146.

Brundin, J. (1964). A functional block in the isthmus of the rabbit fallopian tube. *Acta Physiol. Scand.* **60,** 295–296.

Bryan, J. H. D. (1974). Capacitation in the mouse: The response of murine acrosomes to the environment of the female reproductive tract. *Biol. Reprod.* **10,** 414–421.

Bryan, J. H. D., and Unnithan, R. R. (1972). Non-specific esterase activity in bovine acrosomes. *Histochem. J.* **4,** 413–419.

Burgos, M. H., Sacerdote, F. L., and Russo, J. (1973). Mechanism of sperm release. In *The Regulation of Mammalian Reproductions* (S. J. Segal, R. Crozier, P. A. Corfman, and P. G. Condliffe, eds.), pp. 166–182. Charles C Thomas, Springfield, Ill.

Calarco, P. G., and Szollosi, D. (1973). Intracisternal A particles in ova and preimplantation stages of the mouse. *Nature New Biol.* **243,** 91–93.

Carroll, E. J., and Epel, D. (1975). Isolation and biological activity of the proteases released by sea urchin eggs following fertilization. *Dev. Biol.* **44,** 22–32.

Carter, H. W. (1974). Cumulus cells of the hamster ovum and their interaction with spermatozoa: A correlative light and scanning electron microscopy study. In *Scanning Electron Microscopy/1974*, pp. 623–630. IIT Res. Inst., Chicago.

Chaney, M. O., Demarco, P. V., Jones, N. D., and Occolowitz, J. L. (1974). The structure of A23187, a divalent cation ionophore. *J. Amer. Chem. Soc.* **96,** 1932–1933.

Chang, J. P., and Gibley, C. W. (1968). Ultrastructure of tumor cells during mitosis. *Cancer Res.* **28,** 521–534.

Chang, M. C. (1951). The fertilizing capacity of spermatozoa deposited into the fallopian tubes. *Nature (London)* **168,** 697–698.

Chang, M. C. (1952). Effects of delayed fertilization on segmenting ova, blastocysts and foetuses in rabbits. *Fed. Proc.* **11,** 24.

Chang, M. C. (1954). Development of parthenogenetic rabbit blastocysts induced by low temperature storage of unfertilized ova. *J. Exp. Zool.* **125,** 127–150.

Chang, M. C. (1955). Development of fertilizing capacity of rabbit spermatozoa in the uterus. *Nature (London)* **175,** 1036–1037.

Chang, M. C. (1957). A detrimental effect of seminal plasma on the fertilizing capacity of sperm. *Nature* (*London*) **179,** 258–259.

Chang, M. C. (1959). Fertilization of rabbit ova *in vitro*. *Nature* (*London*) **184,** 466–467.

Chang, M. C. (1966). Effects of oral administration of medoxyprogesterone acetate and ethinyl estradiol on the transportation and development of rabbit eggs. *Endocrinology* **79,** 939–948.

Chang, M. C., and Hunt, D. M. (1956). Effects of proteolytic enzymes on the zona pellucida of fertilized and unfertilized mammalian eggs. *Exp. Cell Res.* **11,** 497–499.

Chase, D. G., and Pikó, L. (1973). Expression of A- and C-type particles in early mouse embryos. *J. Nat. Cancer Inst.* **51,** 1971–1975.

Cholewa-Stewart, J., and Massaro, E. J. (1972). Thermally induced dissolution of the murine zona pellucida. *Biol. Reprod.* **7,** 166–169.

Clermont, Y. (1967). Cinétique de la spermatogenèse chez les mammifères. *Arch. Anat. Microsc.* **56,** 7–60.

Cohen, J., and McNaughton, D. C. (1974). Spermatozoa: The probable selection of a small population by the genital tract of the female rabbit. *J. Reprod. Fertil.* **39,** 297–310.

Collins, F. (1976). A reevaluation of the fertilizing hypothesis of sperm agglutination and the description of a novel form of sperm adhesion. *Dev. Biol.* **49,** 381–394.

Colwin, L. H., and Colwin, A. L. (1967). Membrane fusion in relation to sperm-egg association. In *Fertilization* (C. Metz and A. Monroy, eds.), Vol. 1, pp. 295–367. Academic Press, New York.

Conchie, J., and Mann, T. (1957). Glycosidases in mammalian sperm and seminal plasma. *Nature* (*London*) **179,** 1190–1191.

Cooper, G. W., and Bedford, J. M. (1971). Charge density change in the vitelline surface following fertilization of the rabbit egg. *J. Reprod. Fertil.* **25,** 431–536.

Coppelson, M., and Reid, B. L. (1967). *Preclinical Carcinoma of the Cervix Uteri*. Pergamon Press, Oxford.

Coppelson, M., and Reid, B. L. (1974). Interaction of sperm with somatic cells. *Science* **185,** 239.

Corner, G. W. (1930). The discovery of the mammalian ovum. In *Lectures on the History of Medicine,* 1926–1932. pp. 401–426. W. B. Saunders, Philadelphia.

Courot, M., and Hochereau-de Riviers, M.-T. (1970). Spermatogenesis. In *The Testis* (A. D. Johnson, W. R. Gomes, and N. L. Van Demark, eds.), Vol. 1, Chap. 6, pp. 339–432. Academic Press, New York.

Croce, C. M., Gledhill, B.L., Gabara, B., Sawicki, W., and Koprowski, H. (1972). Lysolecithin-induced fusion of rabbit spermatozoa with hamster somatic cells. In *Workshop on Mechanisms and Prospects of Genetic Exchange* (G. Raspé, ed.), *Advances in the Biosciences,* Vol. 8, pp. 187–200. Pergamon Press, Oxford.

Crogan, D. E., Mayer, D. T., and Sikes, J. D. (1966). Quantitative differences in phospholipids of ejaculated spermatozoa and spermatozoa from three levels of the epididymis of the boar. *J. Reprod. Fertil.* **12,** 431–436.

Cross, P. C., and Brinster, R. L. (1974). Leucine uptake and incorporation at three stages of mouse oocyte maturation. *Exp. Cell Res.* **86,** 43–46.

Csáky, T. Z. (1973). Drugs affecting the exchange of material across cell membranes. In *Fundamentals of Cell Pharmacology* (S. Dikstein, ed.), pp. 322–348. Charles C Thomas, Springfield.

Dan, J. C. (1967). Acrosome reaction and Lysins. In *Fertilization* (C. B. Metz and A. Monroy, eds.), Vol. 1, pp. 237–293. *Academic Press,* New York.

Dauzier, L., Thibault, C., and Wintenberger, S. (1954). La fécondation *in vitro* de l'oeuf de lapine. *C. R. Acad. Sci.* **238,** 844–845.

Dauzier, L., and Wintenberger, S. (1952). La vitesse de remontée des spermatozoïdes dans le tractus génital de la brebis. *Ann. Inst. Nat. Rec. Agron.* **1,** 13–22.

Davis, B. K. (1971). Macromolecular inhibitor of fertilization in rabbit seminal plasma. *Proc. Nat. Acad. Sci.* **68,** 951–955.

Dawson, R. M. C., and Scott, T. W. (1964). Phospholipid composition of epididymal spermatozoa prepared by density gradient centrifugation. *Nature (London)* **202,** 292–293.

de Graaf, R. (1672). *De Mulierum Organis Generatione Inservientibus.*

Doak, R. L., Hall, A., and Dale, H. E. (1967). Longevity of spermatozoa in the reproductive tract of the bitch. *J. Reprod. Fertil.* **13,** 51–58.

Dott, H. M., and Dingle, J. T. (1968). Distribution of lysosomal enzymes in the spermatozoa and cytoplasmic droplets of bull and ram. *Exp. Cell Res.* **52,** 523–540.

Douglas, W. W. (1974). Mechanisms of release of neuro-hypophysial hormones: Stimulus-secretion coupling. In *Handbook of Physiology,* Vol. 4, Sect. 7, pp. 191–224. Amer. Physiol. Soc., Washington.

Dudkiewicz, A. B., Noske, I. G., and Shivers, C. A. (1975). Inhibition of implantation in the golden hamster by zona-precipitating antibody. *Fertil. Steril.* **26,** 686–694.

Dudkiewicz, A. B., Shivers, C. A., and Williams, W. L. (1976). Ultrastructure of hamster zona pellucida treated with zona-precipitating antibody. *Biol. Reprod.* **14,** 175–185.

Dukelow, W. R., Chernoff, H. N., and Williams, W. L. (1966). Enzymatic characterization of decapacitating factor. *Proc. Soc. Exp. Biol. Med.* **121,** 396–398.

Dukelow, W. R., Chernoff, H. N., and Williams, W. L. (1967). Properties of decapacitation factor and presence in various species. *J. Reprod. Fertil.* **14,** 393–399.

Edelman, G. M., and Millette, C. F. (1971). Molecular probes of spermatozoan structures. *Proc. Nat. Acad. Sci.* **68,** 2436–2440.

Edelman, G. M., and Millette, C. (1975). Chemical dissection and surface mapping of spermatozoa. In *The Functional Anatomy of the Spermatozoon* (B. A. Afzelius, ed.), pp. 349–357. Pergamon Press, Oxford.

Edwards, R. G. (1962). Meiosis in ovarian oocytes of adult mammals. *Nature* (*London*) **196,** 446–450.

Edwards, R. G., Bavister, B. D., and Steptoe, P. C. (1969). Early stages of fertilization *in vitro* of human oocytes matured *in vitro*. *Nature* (*London*) **221,** 632–635.

Edwards, R. G., Steptoe, P. C., and Purdy, J. M. (1970). Fertilization and cleavage *in vitro* of preovulatory human oocytes. *Nature* (*London*) **227,** 1307–1309.

Elinson, R. P. (1971). Fertilization of partially jellied and jellyless oocytes of the frog *Rana pipiens*. *J. Exp. Zool.* **176,** 415–428.

Epel, D. (1967). Protein synthesis in sea urchin eggs: A "late" response to fertilization. *Proc. Nat. Acad. Sci.* **57,** 899–906.

Epel, D. (1972). Activation of a Na^+-dependent amino acid transport system upon fertilization of sea urchin eggs. *Exp. Cell Res.* **72,** 74–87.

Epel, D. (1975). The program of and mechanisms of fertilization in the echinoderm egg. *Amer. Zool.* **15,** 507–522.

Epel, D., and Johnson, J. D. (1976). Reorganization of the sea urchin egg surface at fertilization and its relevance to the activation of development. In *Biogenesis and Turnover of Membrane Macromolecules* (J. S.Cook, ed.), pp. 105–120, Raven Press, New York.

Epel, D., Weaver, A. M., Muchmore, A. V., and Schimke, R. T. (1969). β-1,3-Glucanase of sea urchin eggs: Release from particles at fertilization. *Science* **163,** 294–296.

Epel, D., Steinhardt, R., Humphreys, T., and Mazia, D. (1974). An analysis of the partial derepression of sea urchin eggs by ammonia: The existence of independent pathways. *Dev. Biol.* **40,** 245–255.

Ericsson, R. J., Buthala, D. A., and Norland, J. F. (1971). Fertilization of rabbit ova *in vitro* by sperm with adsorbed Sendai virus. *Nature* (*London*) **173,** 54–296.

Espey, L. L. (1967). Ultrastructure of the apex of the rabbit Graffian follicle during the ovulatory process. *Endocrinology* **81,** 267–276.

Evans, E. I. (1933). The transport of spermatozoa in the dog. *Amer. J. Physiol.* **105,** 287–293.

Evans, H. M., and Cole, H. H. (1931). An introduction to the study of the oestrus cycle in the dog. *Mem. Univ. Calif.* **9,** 65–118.

Farris, E. J. (1950). *Human Fertility and Problems of the Male.* Authors Press, White Plains, N.Y.

Fawcett, D. W. (1970). A comparative view of sperm ultrastructure. *Biol. Reprod., Suppl.* **2,** 90–127.

Fawcett, D. W. (1975). Morphogenesis of the mammalian sperm acrosome in new perspective. In *The Functional Anatomy of the Spermatozoon* (B. A. Afzelius, ed.), pp. 199–210. Pergamon Press, Oxford.

First, N. L., Short, R. E., Peters, J. B., and Stratman, F. W. (1965). Transport of spermatozoa in estrual and luteal sows. *J. Anim. Sci.* **24,** 917.

Fléchon, J. -E. (1970). Nature glycoproteique des granules corticaux de l'oeuf de lapine. *J. Microsc. (Paris)* **9,** 221–242.

Fléchon, J. -E., and Dubois, M. P. (1975). Localisation immunocytochimique de la hyaluronidase dans les spermatozoïdes de mammiferes domestique. *C. R. Acad. Sci., Ser. D.*, **280,** 877–880.

Florey, H., and Walton, A. (1932). Uterine fistula used to determine the mechanism of ascent of the spermatozoon in the female genital tract. *J. Physiol.* **74,** 5P.

Fox, L. L., and Shivers, C. A. (1975). Immunologic evidence for addition of oviductal components to the hamster zona pellucida. *Fertil. Steril.* **26,** 599–608.

Franchi, L. L., Mandl, A. M., and Zuckerman, S. (1962). The development of the ovary and the process of oogenesis. In *The Ovary* (S. Zuckerman, ed.), Vol. 1, Chap. 1, pp. 1–88. Academic Press, New York.

Franklin, L. E. (1965). Morphology of gamete fusion and of sperm entry into oocytes of the sea urchin. *J. Cell Biol.* **25,** 81–100.

Franklin, L. E., Barros, C., and Fussell, E. N. (1970). The acrosomal region and the acrosome reaction in sperm of the golden hamster. *Biol. Reprod.* **3,** 180–200.

Fraser, L. R., and Dandekar, P. V. (1975). The relationship between zona digestion and cortical granule disappearance in rabbit eggs inseminated *in vitro*. *Biol. Reprod.* **13,** 123–125.

Fraser, L. R., and Drury, L. M. (1975). The relationship between sperm concentration and fertilization *in vitro* of mouse eggs. *Biol. Reprod.* **13,** 513–518.

Friend, D. S., and Fawcett, D. W. (1974). Membrane differentiations in freeze-fractured mammalian sperm. *J. Cell Biol.* **63,** 641–664.

Fritz, H., Schiessler, H., Schill, W. -B., Tschesche, H., Heimburger, N., and Walker, O. (1975a). Low molecular weight proteinase (acrosin) inhibitors from human and boar seminal plasma and spermatozoa and human cervical mucus—Isolation, properties and biological aspects. In *Proteases and Biological Control*, pp. 737–766. Cold Spring Harbor Laboratory.

Fritz, H., Schleuning, W. -D., Schiessler, H., Schill, W. -B., Wendt, V., and Winkler, G. (1975b). Boar, bull and human sperm acrosin—Isolation, properties and biological aspects. In *Proteases and Biological Control*, pp. 715–735. Cold Spring Harbor Laboratory.

Frye, L. D., and Edidin, M. (1970). The rapid inter-mixing of cell surface antigens after formation of mouse-human heterokaryons. *J. Cell Sci.* **7,** 319–335.

Gaddum-Rosse, P., and Blandau, R. J. (1972). Comparative studies on the proteolysis of fixed gelatin membranes by mammalian sperm acrosomes. *Amer. J. Anat.* **134,** 133–144.

Gall, W. E., Millette, C. F., and Edelman, G. M. (1974). Chemical and structural analysis of mammalian spermatozoa. In *Physiology and Genetics of Reproduction* (E. M. Coutinho and F. Fuchs, eds.), Part A, pp. 241–257. Plenum Press, New York.

Garavagno, A., Posada, J., Barros, C., and Shivers, C. A. (1974). Some characteristics of the zona pellucida antigen. *J. Exp. Zool.* **189,** 37–50.

Garbers, D. L., Wakabayashi, T., and Reed, P. W. (1970). Enzyme profile of the cytoplasmic droplet from bovine epididymal spermatozoa. *Biol. Reprod.* **3,** 327–337.

Garner, D. L., Easton, M. P., Munson, M. E., and Doane, M. A. (1975). Immunofluorescent localization of bovine acrosin. *J. Exp. Zool.* **191,** 127–131.

Gasic, G. J., Berwick, L., and Sorrentino, M. (1968). Positive and negative colloidal iron as cell surface electron stains. *Lab. Invest.* **18,** 63–71.

Gier, H. T., and Marion, G. B. (1970). Development of the mammalian testis. In *The Testis* (A. D. Johnson, W. R. Gomes, and N. L. VanDemark, eds.), Vol. 1, Chap. 1, pp. 1–45. Academic Press, New York.

Gilboa, E., Elkana, Y., and Rigbi, M. (1973). Purification and properties of human acrosin. *Eur. J. Biochem.* **39,** 85–92.

Glass, L. E. (1970). Transmission of maternal proteins into oocytes. *Adv. Biosci.* **6,** 29–58.

Glass, L. E., and Hanson, J. E. (1974). An immunologic approach to contraception: Localization of antiembryo and antizona pellucida serum during mouse preimplantation development. *Fertil. Steril.* **25,** 484–493.

Gledhill, B. L., Gledhill, M. P., Rigler, R., and Ringertz, N. R. (1966). Changes in deoxyribonucleoprotein during spermiogenesis in the bull. *Exp. Cell Res.* **41,** 652–665.

Gordon, M., Dandekar, P. V., and Bartoszewicz, W. (1974). Ultrastructural localization of surface receptors for Concanavalin A on rabbit spermatozoa. *J. Reprod. Fertil.* **36,** 211–214.

Gordon, M., Dandekar, P. V., and Bartoszewicz, W. (1975a). The surface coat of epididymal, ejaculated and capacitated sperm. *J. Ultrastruct. Res.* **50,** 199–207.

Gordon, M., Fraser, L. R., and Dandekar, P. V. (1975b). The effect of ruthenium red and Concanavalin A on the vitelline surface of fertilized and unfertilized rabbit ova. *Anat. Rec.* **181,** 95–112.

Gould, S. F., and Bernstein, M. H. (1975). Localization of bovine sperm hyaluronidase. *Differentiation* **3,** 123–132.

Gould, K., Zaneveld, L. J. D., Srivastava, P. N., and Williams, W. L. (1971). Biochemical changes in the zona pellucida of rabbit ova induced by fertilization and sperm enzymes. *Proc. Soc. Exp. Biol. Med.* **136,** 6–10.

Gould, K. G., Cline, E. M., and Williams, W. L. (1973). Observations on the induction of ovulation and fertilization *in vitro* in the squirrel monkey (*Saimiri sciureus*). *Fertil. Steril.* **24,** 260–268.

Graham, C. F. (1970). Parthenogenetic mouse blastocysts. *Nature (London)* **226,** 165–167.

Graham, C. F. (1971). Experimental early parthenogenesis in mammals. *Adv. Biosci.* **6,** 87–97.

Graham, C. F. (1974). The production of parthenogenetic mammalian embryos and their use in biological research. *Biol. Rev.* **49,** 399–422.

Greenslade, F. C., McCormack, J. J., Hirsch, A. F., and Danvanzo, J. P. (1973). Blockage of fertilization in *Rana pipiens* by trypsin inhibitors. *Biol. Reprod.* **8,** 306–310.

Gregory, P. W. (1930). The early embryology of the rabbit. *Contr. Embryol. Carnegie Inst.* **21,** 141–168.

Gresson, R. A. R. (1941). A study of the cytoplasmic inclusions during maturation, fertilization and the first cleavage division of the egg of the mouse. *Quart. J. Micro. Sci.* **83,** 35–59.

Gulyas, B. J. (1974). Cortical granules in artificially activated (parthenogenetic) rabbit eggs. *Amer. J. Anat.* **140,** 577–582.

Gurdon, J. B., and Woodland, H. R. (1968). The cytoplasmic control of nuclear activity in animal development. *Biol. Rev.* **43,** 233–267.

Gwatkin, R. B. L. (1967). Passage of mengovirus through the zona pellucida of the mouse morula. *J. Reprod. Fertil.* **13,** 577–578.

Gwatkin, R. B. L. (1975). Recent advances in gamete biology and their possible applications to fertility control. *Ann. Reports Med. Chem.* **10,** 240–245.

Gwatkin, R. B. L. (1976). Fertilization. In *The Cell Surface in Embryogenesis and Development* (G. Poste, and G. L. Nicolson, eds.), Chap. 1, pp. 1–54. North-Holland, Amsterdam.

Gwatkin, R. B. L., and Andersen, O. F. (1969). Capacitation of hamster spermatozoa by bovine follicular fluid. *Nature (London)* **224,** 1111–1112.

Gwatkin, R. B. L., and Andersen, O. F. (1973). Effect of glycosidase inhibitors on the capacitation of hamster spermatozoa by cumulus cells *in vitro*. *J. Reprod. Fertil.* **35,** 565–567.

Gwatkin, R. B. L. and Andersen, O. F. (1976). Hamster oocyte maturation *in vitro:* Inhibition by follicular components. *Life Sci.* **19,** 527–536.

Gwatkin, R. B. L., and Carter, H. E. (1975). Cumulus oophorus. In *Scanning Electron Microscope Atlas of Mammalian Reproduction* (E. S. E. Hafez, ed.). Igaku-Shoin, Tokyo and Springer-Verlag, New York.

Gwatkin, R. B. L., and Haidri, A. A. (1973). Requirements for the maturation of hamster oocytes *in vitro*. *Exp. Cell Res.* **76,** 1–7.

Gwatkin, R. B. L., and Hutchison, C. F. (1971). Capacitation of hamster spermatozoa by β-glucuronidase. *Nature (London)* **229,** 343–344.

Gwatkin, R. B. L., and Williams, D. T. (1970). Inhibition of sperm capacitation *in vitro* by contraceptive steroids. *Nature (London)* **227,** 182–183.

Gwatkin, R. B. L., and Williams, D. T. (1974). Heat sensitivity of the cortical granule protease from hamster eggs. *J. Reprod. Fertil.* **39,** 153–155.

Gwatkin, R. B. L., and Williams, D. T. (1976a). Receptor activity of the solubilized zona pellucida. Ninth Ann. Meeting, Soc. Study Reprod., Philadelphia.

Gwatkin, R. B. L., and Williams, D. T. (1976b). Receptor activity of the solubilized hamster and mouse zona pellucida before and after the zona reaction. *J. Reprod. Fertil.* **49,** 55–59.

Gwatkin, R. B. L., Andersen, O. F., and Hutchison, C. F. (1972). Capacitation of hamster spermatozoa *in vitro*: The role of cumulus components. *J. Reprod. Fertil.* **30,** 389–394.

Gwatkin, R. B. L., Andersen, O. F. , and Williams, D. T. (1974). Capacitation of mouse spermatozoa *in vitro*: Involvement of epididymal secretions and cumulus oophorus. *J. Reprod. Fertil.* **41,** 253–256.

Gwatkin, R. B. L., Williams, D. T., and Andersen, O. F. (1973a). Zona reaction of mammalian eggs: Properties of the cortical granule protease (Cortin) and its receptor substrate in hamster eggs. *J. Cell Biol.* **59,** 128a.

Gwatkin, R. B. L., Williams, D. T., Hartmann, J. F., and Kniazuk, M. (1973b). The zona reaction of hamster and mouse eggs: Production *in vitro* by a trypsin-like protease from cortical granules. *J. Reprod. Fertil.* **32,** 259–265.

Gwatkin, R. B. L., Carter, H. W., and Patterson, H. (1976a). Association of mammalian sperm with the cumulus cells and the zona pellucida studied by scanning electron microscopy. In *Scanning Electron Microscopy* (H. Johari and R. P. Becker, eds.), Vol. 2, Part 2, pp. 379–384. IIT Res. Inst., Chicago, Ill.

Gwatkin, R. B. L., Rasmusson, G. H., and Williams, D. T. (1976b). Induction of the cortical reaction in hamster eggs by membrane-active agents. *J. Reprod. Fertil.* **47,** 299–303.

Halbert, S. A., Tam, P. Y., and Blandau, R. J. (1976). Egg transport in the rabbit oviduct: The role of cilia and muscle. *Science* **191,** 1052–1053.

Ham, R. G. (1963). An improved nutrient solution for diploid Chinese hamster and human cell lines. *Exp. Cell Res.* **29,** 515–526.

Ham, R. G. (1965). Clonal growth of mammalian cells in a chemically defined synthetic medium. *Prod. Nat. Acad. Sci.* **53,** 288–293.

Hamilton, W. J., and Day, F. T. (1945). Cleavage stages of the ova of the horse with notes on ovulation. *J. Anat.* **79,** 127–130.

Hammond, J., and Asdell, S. A. (1926). The vitality of spermatozoa in the male and female reproductive tract. *J. Exp. Biol.* **4,** 155–185.

Hamner, C. E. (1973). Physiology of sperm in the female reproductive tract. In *The Regulation of Mammalian Reproduction* (S. J. Segal, R. Crozier, P. A. Corfman, and P. G. Condliffe, eds.), pp. 203–212. Charles C Thomas, Springfield, Ill.

Hamner, C. E., Jennings, L. L., and Sojka, N. J. (1970). Cat (*Felis catus*) spermatozoa require capacitation. *J. Reprod. Fertil.* **23,** 477–480.

Hanada, A., and Chang, M. C. (1972). Penetration of zona-free eggs by spermatozoa of different species. *Biol. Reprod.* **6,** 300–309.

Hanada, A., and Chang, M. C. (1976). Penetration of hamster and rabbit zona-free eggs by rat and mouse spermatozoa with special reference to sperm capacitation. *J. Reprod. Fertil.* **46,** 239–241.

Harper, M. J. K. (1973). Stimulation of sperm movement from the isthmus to the site of fertilization in the rabbit oviduct. *Biol. Reprod.* **8,** 369–377.

Hartman, C. G. (1916). Studies on the development of the opossum (*Didelphis virginiana* L.). I. History of early cleavage. II. Formation of the blastocyst. *J. Morphol.* **27,** 1–83.

Hartman, C. G. (1939). Ovulation and the transport and viability of ova and sperm in the female genital tract. In *Sex and Internal Secretions*, 2nd Ed. (E. Allen, ed.), Williams & Wilkins, Baltimore.

Hartmann, J. F., and Gwatkin, R. B. L. (1971). Alteration of sites on the mammalian sperm surface following capacitation. *Nature (London)* **234,** 479–481.

Hartmann, J. F., and Hutchison, C. F. (1974a). Nature of the prepenetration contact interactions between hamster gametes *in vitro*. *J. Reprod. Fertil.* **36,** 49–57.

Hartmann, J. F., and Hutchison, C. F. (1974b). Contact between hamster spermatozoa and the zona pellucida releases a factor which influences early binding stages. *J. Reprod. Fertil.* **37,** 61–66.

Hartmann, J. F., and Hutchison, C. F. (1974c). Mammalian fertilization *in vitro*: Sperm induced preparation of the zona pellucida of golden hamster ova for final binding. *J. Reprod. Fertil.* **37,** 443–445.

Hartmann, J. F., and Hutchison, C. F. (1975). The effect of sperm concentration on binding to and penetration of hamster eggs *in vitro*. *Abstract No. 50, Eighth Annual Meeting, Soc. Study of Reprod.*, Fort Collins, Colo.

Hartmann, J. F., and Hutchison, C. F. (1976). Is acrosin the lysin of the zona pellucida during fertilization? *Abstract No. 2, Ninth Annual Meeting, Soc. Study of Reprod.*, Philadelphia.

Hartmann, J. F., Gwatkin, R. B. L., and Hutchison, C. F. (1972). Early contact interactions between mammalian gametes *in vitro*: Evidence that the vitellus influences adherence between sperm and zona pellucida. *Proc. Nat. Acad. Sci.* **69,** 2767–2769.

Hartree, A. C. (1975). The acrosome–lysosome relationship. *J. Reprod. Fertil.* **44,** 125–126.

Harvey, W. (1651). *De Generatione Animalium*, London.

Hastings, R. A., Enders, A. C., and Schlafke, S. (1972). Permeability of the zona pellucida to protein tracers. *Biol. Reprod.* **7,** 288–296.

Henle, W., Henle, G., and Chambers, L. A. (1938). Studies on the antigenic structure of some mammalian spermatozoa. *J. Exp. Med.* **68,** 335–352.

Hertwig, O. (1876). Beiträge zur Kenntniss der Bildung, Befruchtung und Theilung des tierischen Eies. *Gegenbaurs Morph. Jb.* **1,** 347–434.

Higgins, P. J., Borenfreund, E., and Bendich, A. (1975). Appearance of foetal antigens in somatic cells after interaction with heterologous sperm. *Nature (London)* **257,** 488–489.

Hjort, T., and Brogaard, K. (1971). The detection of different spermatozoal antibodies and their occurrence in normal and infertile women. *Clin. Exp. Immunol.* **8,** 9–23.

Hodgson, B. J., and Pauerstein, C. J. (1975). Effect of hormonal treatments which alter ovum transport on β-adrenoreceptors of the rabbit oviduct. *Fertil. Steril.* **26,** 573–578.

Hope, J. (1965). The fine structure of the developing follicle of the rhesus ovary. *J. Ultrastruct. Res.* **12,** 592–610.

Hoppe, P. C., and Pitts, S. (1973). Fertilization *in vitro* and development of mouse ova. *Biol. Reprod.* **8,** 420–426.

Hoppe, P. C., and Whitten, W. K. (1974). An albumin requirement for fertilization of mouse eggs *in vitro*. *J. Reprod. Fertil.* **39,** 433–436.

Howe, G. R. (1967). Leucocytic response to spermatozoa in ligated segments of the rabbit vagina, uterus and oviduct. *J. Reprod. Fertil.* **13,** 563–566.

Howe, G. R., and Black, D. L. (1963). Spermatozoan transport and leucocytic responses in the reproductive tract of calves. *J. Reprod. Fertil.* **6,** 305–311.

Hunter, R. H. F. (1967). The effects of delayed insemination on fertilization and early cleavage in the pig. *J. Reprod. Fertil.* **13,** 133–147.

Hunter, R. H. F., and Hall, J. P. (1974). Capacitation of boar spermatozoa: Synergism between uterine and tubal environments. *J. Exp. Zool.* **188,** 203–214.

Hunter, M., and Nornes, H. O. (1969). Characterization and isolation of a sperm-coating antigen from rabbit seminal plasma with capacity to block fertilization. *J. Reprod. Fertil.* **20,** 419–427.

Iles, S. A., McBurney, M. W., Bramwell, S. R., Deussen, Z. A. and Graham, C. F. (1975). Development of parthenogenetic and fertilized mouse embryos in the uterus and in extrauterine sites. *J. Embryol. Exp. Morphol.* **34,** 387–405.

Inoue, M. (1973). Physiochemical characterization of the murine zona pellucida. *Biol. Reprod.* **9,** 80.

Inoue, M., and Wolf, D. P. (1974). Comparative solubility properties of the zonae pellucidae of unfertilized and fertilized mouse ova. *Biol. Reprod.* **11,** 558–565.

Inoue, M., and Wolf, D. P. (1975). Fertilization-associated changes in the murine zona pellucida: A time sequence study. *Biol. Reprod.* **13,** 546–551.

Iwamatsu, T., and Chang, M. C. (1971). Factors involved in the fertilization of mouse eggs *in vitro. J. Reprod. Fertil.* **26,** 197–208.

Iwamatsu, T., and Chang, M. C. (1972). Sperm penetration *in vitro* of mouse oocytes at various times during maturation. *J. Reprod. Fertil.* **31,** 237–247.

Jacoby, F. (1962). Ovarian histochemistry. In *The Ovary* (S. Zuckerman, ed.), Vol. 1, pp. 189–245. Academic Press, New York.

Ji, T. H., and Nicolson, G. L. (1974). Lectin binding and perturbation of the outer surface of the cell membrane induces a trans-membrane organizational alteration at the inner surface. *Proc. Nat. Acad. Sci.* **71,** 2212–2216.

Jilek, F., and Pavlok, A. (1975). Antibodies against mouse ovaries and their effect on fertilization *in vitro* and *in vitro* in the mouse. *J. Reprod. Fertil.* **42,** 377–380.

Johnson, M. H., and Edidin, M. (1972). H-2 antigens on mouse spermatozoa. *Transplantation* **14,** 781–786.

Johnson, J. D., and Epel, D. (1075). Relationship between release of surface proteins and metabolic activation of sea urchin eggs at fertilization. *Proc. Nat. Acad. Sci.* **72,** 4474–4478.

Johnson, M. H., Eager, D., Muggleton-Harris, A., and Grave, H. M. (1975). Mosaicism in organization of Concanavalin A receptors on surface membrane of mouse egg. *Nature (London)* **257,** 321–322.

Johnson, W. L., and Hunter, A. G. (1972). Seminal antigens: Their alteration in the genital tract of female rabbits and during partial *in vitro* capacitation with beta amylase and beta glucuronidase. *Biol. Reprod.* **7,** 332–340.

Jones, R. C. (1973). Changes occurring in the head of boar spermatozoa: Vesiculation or vacuolation of the acrosome? *J. Reprod. Fertil.* **33,** 113–118.

Kaufman, M. H., and Surani, M. A. H. (1974). The effect of osmolarity on mouse parthenogenesis. *J. Embryol. Exp. Morphol.* **31,** 513–526.

Killian, G. J., and Amann, R. P. (1973). Immunoelectrophoretic characterization of fluid and sperm entering and leaving the bovine epididymis. *Biol. Reprod.* **9,** 489–499.

Kirton, K. T., and Hafs, H. D. (1965). Sperm capacitation by uterine fluid or beta-amylase *in vitro. Science* **150,** 618–619.

Koehler, J. K., and Gaddum-Rosse, P. (1975). Media induced alterations of the membrane associated particles of the guinea pig sperm tail. *J. Ultrastruct. Res.* **51,** 106–118.

Komar, A. (1973). Parthenogenetic development of mouse eggs activated by heat shock. *J. Reprod. Fertil.* **35,** 433–443.

Kopecny, V., and Pavlok, A. (1975). Autoradiographic study of mouse spermatozoan arginine-rich nuclear protein in fertilization. *J. Exp. Zool.* **191,** 85–96.

Koren, E., and Milkovic, S. (1973). Collagenase-like peptidase in human, rat and bull spermatozoa. *J. Reprod. Fertil.* **32,** 349–356.

Krzanowska, H. (1972). Rapidity of removal *in vitro* of the cumulus oophorus and the zona pellucida in different strains of mice. *J. Reprod. Fertil.* **31,** 7–14.

Kuehl, T. J., and Dukelow, W. R. (1975). Squirrel monkey follicular oocyte recovery, *in vitro* fertilization and embryo transfer via the laparoscope. *Abstract No. 79, Eighth Annual Meeting, Soc. Study Reprod.*, Fort Collins, Colo.

Laing, J. A. (1945). Observations on the survival time of the spermatozoa in the genital tract of the cow and its relation to fertility. *J. Agric. Sci.* **35,** 72–83.

Lallier, R. A. (1970). Formation of fertilization membrane in sea urchin eggs. *Exp. Cell Res.* **63,** 460–462.

Lallier, R. A. (1971). Effects of various inhibitors of protein cross-linking on the formation of fertilization membrane in sea urchin egg. *Experientia* **27,** 1323–1324.

Lams, H. (1913). Etude de l'oeuf de cobaye aux premiers stades de l'embryogenèse. *Arch. Biol.* (*Paris*) **28,** 229–323.

Lewis, B. K., and Ketchell, M. M. (1972). Effects of female reproductive tract secretions on rabbit sperm. I. Release of hyaluronidase *in vitro. Proc. Soc. Exp. Biol. Med.*, **141,** 712–718.

Lewis, W. H., and Hartman, C. G. (1941). Tubal ova of the rhesus monkey. *Contr. Embryol. Carnegie Inst.* **29,** 1–14.

Lewis, W. H., and Wright, E. S. (1935). On the early development of the mouse egg. *Contr. Embryol. Carnegie Inst.* **25,** 115–143.

Loewenstein, J. E., and Cohen, A. I. (1964). Dry mass, lipid content and protein content of the intact and zona-free mouse zona. *J. Embryol. Exp. Morph.* **12,** 113–121.

Long, J. A. (1912). The living eggs of rats and mice with a description of apparatus for obtaining and observing them. *Univ. Calif. Publ. Zool.* **9,** 105–136.

Longo, F. J. (1973). Fertilization: A comparative ultrastructural review. *Biol. Reprod.* **9,** 149–215.

Longo, F. J. (1975). Ultrastructural analysis of artificially activated rabbit eggs. *J. Exp. Zool.* **192,** 87–112.

Longo, F. J., Schuel, H., and Wilson, W. L. (1975). Mechanism of soybean trypsin inhibitor induced polyspermy as determined by an analysis of refertilized sea urchin (*Arbacia punctulata*) eggs. *Dev. Biol.* **41,** 193–201.

Lopata, A., Johnston, I. W. H., Leeton, J. F., Muchnicki, D., McTalbot, J., and Wood, C. (1974). Collection of human oocytes at laparoscopy and laparotomy. *Fertil. Steril.* **25,** 1030–1038.

Lord Rothschild (1956a). *Fertilization.* Methuen & Co., London.

Lord Rothschild (1956b). Unorthodox methods of sperm transfer. *Sci. Amer.* **195,** 121–132.

Lowe, D. A., Richardson, B. P., Taylor, P., and Donatsch, P. (1976). Increasing intracellular sodium triggers calcium release from bound pools. *Nature (London)* **260,** 337–338.

Lung, B. (1974). Architecture of mammalian sperm: Analysis by quantitative electron microscopy. *Adv. Cell. Molec. Biol.* **3,** 73–133.

Luthardt, F. W., and Donahue, R. P. (1973). Pronuclear DNA synthesis in mouse eggs. *Exp. Cell Res.* **82,** 143–151.

Mahajan, S. C., and Menge, A. C. (1966). Factors influencing the disposal of sperm and the leukocytic response in the rabbit uterus. *Int. J. Fertil.* **11,** 373–380.

Mahi, C. A. (1975). Maturation and fertilization *in vitro* of canine ovarian oocytes. *Abstract No. 78, Eighth Annual Meeting, Soc. Study Reprod.*, Fort Collins, Colo.

Mahi, C. A., and Yanagimachi, R. (1973). The effects of temperature, osmolarity and hydrogen ion concentration on the activation and acrosome reaction of golden hamster spermatozoa. *J. Reprod. Fertil.* **35,** 55–56.

Mahi, C. A., and Yanagimachi, R. (1975). Induction of nuclear decondensation of mammalian spermatozoa *in vitro. J. Reprod. Fertil.* **44,** 293–296.

Mainland, D. (1930). The early development of the ferret: The pronuclei. *J. Anat.* **64,** 262–287.

Malpighi, M. (1673). *De Formatione Pulli In Ovo*, Bologna.

Mann, T. (1964). *The biochemistry of semen and of the male reproductive tract.* Methuen & Co., London.

Mann, T. (1967). Sperm metabolism. In *Fertilization* (C. B. Metz and A. Monroy, eds.), Vol. 1, pp. 99–116. Academic Press, New York.

Mann, T., Polge, C., and Rowson, L. E. A. (1956). Participation of seminal plasma during the passage of spermatozoa in the female reproductive tract of the pig and horse. *J. Endocrinol.* **13,** 133–140.

Marston, J. H., and Chang, M. C. (1964). The fertilizable life of ova and their morphology following delayed insemination in mature and immature mice. *J. Exp. Zool.* **155,** 237–252.

Mastroianni, L., and Jones, R. (1964). Oxygen tension within the rabbit fallopian tube. *J. Reprod. Fertil.* **9,** 99–102.

Masui, Y. (1974). A cytostatic factor in amphibian oocytes: Its extraction and partial characterization. *J. Exp. Zool.* **187,** 141–147.

Masui, Y., and Markert, C. L. (1971). Cytoplasmic control of nuclear behavior during meiotic maturation of frog oocytes. *J. Exp. Zool.* **177,** 129–146.

Mattner, P. E., and Braden, A. W. H. (1963). Spermatozoa in the genital tract of the ewe. I. Rapidity of transport. *Austral. J. Biol. Sci.* **16,** 473–481.

McReynolds, H. D., and Hadek, R. (1971). A study on sperm tail elements in mouse blastocysts. *J. Reprod. Fertil.* **24,** 291–294.

Meizel, S., and Cotham, J. (1972). Partial characterization of a new bull sperm arylaminidase. *J. Reprod. Fertil.* **28,** 303–307.

Meizel, S., and Lui, C. W. (1976). Evidence for the role of a trypsin-like enzyme in the hamster sperm acrosome reaction. *J. Exp. Zool.* **195,** 137–144.

Meizel, S., and Mukerji, S. K. (1975). Proacrosin from rabbit epididymal spermatozoa: Partial purification and initial biochemical characterization. *Biol. Reprod.* **13,** 83–93.

Merton, H. (1939). Studies on reproduction in the albino mouse. III. The duration of life of spermatozoa in the female reproductive tract. *Proc. Royal Soc. (Edinburgh)* **59,** 207–218.

Metz, C. B. (1972). Effect of antibodies on gametes and fertilization. *Biol. Reprod.* **6,** 358–383.

Metz, C. B. (1973). Role of specific sperm antigens in fertilization. *Fed. Proc.* **32,** 2057–2064.

Metz, C. B., and Monroy, A., eds. (1969). *Fertilization, comparative morphology, biochemistry and immunology.* Academic Press, New York.

Mintz, B., and Gearhart, J. D. (1973). Subnormal zona pellucida changes in parthenogenetic mouse embryos. *Dev. Biol.* **31,** 178–184.

Miyamoto, H., and Chang, M. C. (1972). Development of mouse eggs fertilized *in vitro* by epididymal spermatozoa. *J. Reprod. Fertil.* **30,** 135–137.

Miyamoto, H., and Chang, M. C. (1973a). Effects of protease inhibitors on the fertilizing capacity of hamster spermatozoa. *Biol. Reprod.* **9,** 533–537.

Miyamoto, H., and Chang, M. C. (1973b). The importance of serum albumin and metabolic intermediates for capacitation of spermatozoa and fertilization of mouse eggs *in vitro. J. Reprod. Fertil.* **32,** 193–205.

Miyamoto, H., and Ishibashi, T. (1975). The role of calcium ions in fertilization of mouse and rat eggs *in vitro. J. Reprod. Fertil.* **45,** 523–526.

Miyamoto, H., Toyoda, Y., and Chang, M. C. (1974). Effect of hydrogen ion concentration on *in vitro* fertilization of mouse, golden hamster and rat eggs. *Biol. Reprod.* **10,** 487–493.

Monroy, A. (1965). *Chemistry and Physiology of Fertilization.* Holt, Rhinehart & Winston, New York.

Monroy, A. (1973). Fertilization and its biochemical consequences. *Addison-Wesley Module in Biology*, No. 7. Addison-Wesley, Reading, Mass.

Monroy, A., Ortolani, G., O'Dell, D., and Millonig, G. (1973). Binding of Concanavalin A to the surface of unfertilized and fertilized *Ascidian* eggs. *Nature* (*London*) **242,** 409–410.

Morgan, J. F., Morton, J. J., and Parker, R. C. (1950). Nutrition of animal cells in tissue culture. I. Initial studies on a synthetic medium. *Proc. Soc. Exp. Biol. Med.* **73,** 1.

Morton, D. B. (1975). Acrosomal enzymes: Immunochemical localization of acrosin and hyaluronidase in ram spermatozoa. *J. Reprod. Fertil.* **45,** 375–378.

Morton, D. B., and Bavister, B. D. (1974). Fractionation of hamster sperm-capacitating components from human serum by gel filtration. *J. Reprod. Fertil.* **40,** 491–493.

Moyer, D. L., Legorreta, G., Maruta, H., and Henderson, V. (1967). Elimination of homologous spermatozoa in the female genital tract of the rabbit: A light and electron-microscope study. *J. Pathol. Bacteriol.* **94,** 345–350.

Moyer, D. L., Rimdusit, S. and Mischell, D. R. Jr. (1970). Sperm distribution and degradation in the human female reproductive tract. *Obstet. Gynecol.* **35,** 831–840.

Mukhergee, A. B., and Cohen, M. M. (1970). Development of normal mice by *in vitro* fertilization. *Nature* (*London*) **228,** 472–473.

Murdoch, R. N., and White, I. G. (1967). The metabolism of unlabelled glucose by rabbit spermatozoa after incubation *in utero*. *J. Reprod. Fertil.* **14,** 213–223.

Nakamura, M., and Yasumasu, I. (1974). Mechanism for increase in intracellular concentration of free calcium in fertilized sea urchin egg: A method for estimating intracellular concentration of free calcium. *J. Gen. Physiol.* **63,** 374–388.

Nelson, L. (1967). Sperm motility. In *Fertilization* (C. B. Metz and A. Monroy, eds.), Vol. 1, pp. 27–97. Academic Press, New York.

Nicolson, G. L. (1973). The relationship of a fluid membrane structure to cell agglutination and surface topography. *Ser. Haemat.* **6,** 275–291.

Nicolson, G. L., and Yanagimachi, R. (1972). Terminal saccharides on sperm plasma membranes: Identification by specific agglutinins. *Science* **177,** 276–279.

Nicolson, G. L., and Yanagimachi, R. (1974). Mobility and restriction of mobility of plasma membrane lectin-binding components. *Science* **184,** 1294–1296.

Nicolson, G. L., Yanagimachi, R., and Yanagimachi, H. (1975). Ultrastructural localization of lectin binding sites on the zona pellucida and plasma membranes of mammalian eggs. *J. Cell Biol.* **66,** 263–274.

Nicosia, S. B., and Mikhail, G. (1975). Cumuli oophori in tissue culture: Hormone production, ultrastructure and morphometry of early luteinization. *Fertil. Steril.* **26,** 427–448.

Niwa, K., and Chang, M. C. (1975). Requirement of capacitation for sperm pentration of zona-free rat eggs. *J. Reprod. Fert.* **44,** 305–308.

Norrevang, A. (1968). Electron microscopic morphology of oogenesis. *Int. Rev. Cytol.* **23,** 113–186.

Oakberg, E. F., and Tyrrell, P. D. (1975). Labeling the zona pellucida of the mouse oocyte. *Biol. Reprod.* **12,** 477–482.

Odor, D. L. (1960). Electron microscopic studies on ovarian oocytes and unfertilized tubal ova in the rat. *J. Biophys. Biochem. Cytol.* **7,** 567–574.

Ogawa, S., Satoh, K., Hamada, M., and Hashimoto, H. (1972). *In vitro* culture of rabbit ova fertilized by epididymal sperms in chemically defined medium. *Nature* (*London*) **238,** 270–271.

Oh, Y. K., and Brackett, B. G. (1975). Ultrastructure of rabbit ova recovered from ovarian follicles and inseminated *in vitro. Fertil. Steril.* **26,** 665–685.

Oikawa, T., and Yanagimachi, R. (1975). Block of hamster fertilization by anti-ovary antibody. *J. Reprod. Fertil.* **45,** 487–494.

Oikawa, T., Yanagimachi, R., and Nicolson, G. L. (1973). Wheat germ agglutinin blocks mammalian fertilization. *Nature* (*London*) **241,** 256–259.

Oikawa, T., Nicolson, G. L., and Yanagimachi, R. (1974). Inhibition of hamster fertilization by phytoagglutinins. *Exp. Cell Res.* **83,** 239–246.

Oikawa, T., Nicolson, G. L., and Yanagimachi, R. (1975). Trypsin-mediated modification of the zona pellucida glycopeptide structure of hamster eggs. *J. Reprod. Fertil.* **43,** 133–136.

Oliphant, G. (1976). Removal of sperm-bound seminal plasma components as a prerequisite to induction of the rabbit acrosome reaction. *Fertil. Steril.* **27,** 28–38.

Oliphant, G., and Brackett, B. G. (1973a). Immunological assessment of surface changes of rabbit sperm undergoing capacitation. *Biol. Reprod.* **9,** 404–414.

Oliphant, G., and Brackett, B. G. (1973b). Capacitation of mouse spermatozoa in media with elevated ionic strength and reversible decapacitation with epididymal extracts. *Fertil. Steril.* **24,** 948–955.

Olson, G. E., Hamilton, D. W., and Fawcett, D. W. (1976). Isolation and characterization of the perforatorium of rat spermatozoa. *J. Reprod. Fertil.* **47,** 293–297.

O'Rand, M. G. (1972). *In vitro* fertilization and capacitation-like interaction in the hydroid *Campanularia flexuosa. J. Exp. Zool.* **182,** 299–305.

O'Rand, M. G. (1974). Gamete interaction during fertilization in *Campanularia*—The female epithelial cell surface. *Amer. Zool.* **14,** 487–493.

Orgebin-Crist, M. -C. (1969). Studies on the function of the epididymis. *Biol. Reprod. Suppl. 1*, 155–175.

Overstreet, J. W. (1975). Human sperm penetration of non-living human oocytes *in vitro*. *Abstract No. 55*, *Annual Meeting*, *Soc. Study Reprod.*, Fort Collins, Colo.

Overstreet, J. W., and Bedford, J. M. (1974a). Transport, capacitation and fertilizing ability of epididymal spermatozoa. *J. Exp. Zool.* **189,** 203–214.

Overstreet, J. W., and Bedford, J. M. (1974b). Comparison of the penetrability of the egg vestments in follicular oocytes, unfertilized and fertilized ova of the rabbit. *Dev. Biol.* **41,** 185–192.

Overstreet, J. W., and Bedford, J. M. (1975). The penetrability of rabbit ova treated with enzymes or anti-progesterone antibody: A probe into the nature of a mammalian fertilizin. *J. Reprod. Fertil.* **44,** 273–284.

Ownby, C. L., and Shivers, C. A. (1972). Antigens of the hamster ovary and effects of anti-ovary serum on eggs. *Biol. Reprod.* **6,** 310–318.

Ozawa, E. K., Hosoi, K., and Ebashi, S. (1967). Reversible stimulation of muscle phosphorylase *b* kinase by low concentration of calcium ions. *J. Biochem.* **61,** 531.

Pache, W. (1975). Boromycin. In *Antibiotics* (J. W. Corcoran and F. E. Hahn, eds.), Vol. 3, pp. 585–587. Springer-Verlag, Heidelberg.

Papahadjopoulos, D., and Poste, G. (1975). Calcium-induced phase separation and fusion in phospholipid membranes. *Biophys. J.* **15,** 945–948.

Papahadjopoulos, D., Poste, G., Schaeffer, B. E., and Vail, W. J. (1974). Membrane fusion and molecular segregation in phospholipid vesicles. *Biochim. Biophys. Acta* **352,** 10–28.

Papahadjopoulos, D., Vail, W. J., Jacobson, K., and Poste, G. (1975). Cochleate lipid cylinders produced by fusion of unilamellar lipid vesicles. *Biochim. Biophys. Acta* **394,** 483–491.

Parr, E. L. (1975). Rupture of ovarian follicles. *J. Reprod. Fertil.*, *Suppl.* **22,** 1.

Pedersen, H. (1969). Ultrastructure of the ejaculated human sperm. *Z. Zellforsch.* **94,** 542–554.

Penn, A., Gledhill, B. L., and Darzynkiewicz, Z. (1972). Modification of the gelatin substrate procedure for demonstration of acrosomal proteolytic activity. *J. Histchem. Cytochem.* **20,** 499–506.

Pethica, B. A. (1961). The physical chemistry of cell division. *Exp. Cell Res.*, *Suppl.* **8,** 123–140.

Phillips, D. M. (1972). Substructure of the mammalian acrosome. *J. Ultrastruct. Res.* **38,** 591–604.

Phillips, D. M. (1975). Cell surface structure of rodent sperm heads. *J. Exp. Zool.* **191,** 1–8.

Piatigorsky, J., and Whiteley, A. H. (1965). A change in permeability and uptake of ^{14}C-uridine in response to fertilization in *Strongylocentrotus purpuratus* eggs. *Biochim. Biophys. Acta* **108,** 404–418.

Pickworth, S., and Chang, M. C. (1969). Fertilization of Chinese hamster eggs *in vitro*. *J. Reprod. Fertil.* **19,** 371–374.

Pienkowski, M. (1974). Study of the growth regulation of pre-implantation mouse embryos using Concanavalin A. *Proc. Soc. Exp. Biol. Med.* **145,** 464–469.

Pikó, L. (1961). Repeated fertilization of fertilized rat eggs after treatment with versene (EDTA). *Amer. Zool.* **1,** 467–468.

Pikó, L. (1969). Gamete structure and sperm entry in mammals. In *Fertilization* (C. Metz and A. Monroy, eds.), Vol. 2, pp. 325–403. Academic Press, New York.

Pincus, G. (1939). The breeding of some rabbits produced by recipients of artificially activated ova. *Proc. Nat. Acad. Sci.* **25,** 557–559.

Pincus, G., and Enzmann, E. V. (1936). The comparative behavior of mammalian eggs *in vivo* and *in vitro*. II. Activation of tubal eggs of the rabbit. *J. Exp. Zool.* **73,** 195–208.

Pincus, G., and Shapiro, H. (1940). Further studies on the parthenogenetic activation of rabbit eggs. *Proc. Nat. Acad. Sci.* **26,** 163–165.

Pinsker, M. C., and Williams, W. L. (1967). Properties of a spermatozoa anti-fertility factor. *Arch. Biochem. Biophys.* **122,** 111–117.

Pitkjanen, I. G. (1960). The fate of spermatozoa in the uterus of the sow. *Z. Obsc. Biol.* **21,** 28–33.

Polakoski, K. L. (1974). Partial purification and characterization of proacrosin from boar sperm. *Fed. Proc.* **33,** 1308.

Polakoski, K. L., and McRorie, R. A. (1973). Boar acrosin. II. Classification, inhibition and specificity studies of a proteinase from sperm acrosomes. *J. Biol. Chem.* **248,** 8183–8188.

Polakoski, K. L., McRorie, R. A., and Williams, W. L. (1973). Boar acrosin. I. Purification and preliminary characterization of a proteinase from boar sperm acrosomes. *J. Biol. Chem* **248,** 8178–8182.

Poste, G. (1970). Virus-induced polykaryocytosis and the mechanism of cell fusion. *Adv. Virus Res.* **16,** 303–356.

Poste, G. (1972). Mechanisms of virus-induced cell fusion. *Int. Rev. Cytol.* **33,** 157–252.

Poste, G., and Allison, A. C. (1973). Membrane fusion. *Biochem. Biophys. Acta* **300,** 421–465.

Pressman, B. C. (1973). Properties of ionophores with broad range cation selectivity. *Fed. Proc.* **32,** 1698–1704.

Quinn, P. J., and White, I. G. (1967). The phospholipid and cholesterol content of epididymal and ejaculated ram spermatozoa and seminal plasma in relation to cold shock. *Austral. J. Biol. Sci.* **20,** 1205–1215.

Repin, V. S., and Akimova, I. M. (1976). The microelectrophoretic analysis of protein patterns of mammalian oocyte and zygote zonae pellucidae. *Biokimia* (USSR) **41,** 50–57.

Reyes, A., and Rosado, A. (1975). Interference with sperm binding to zona pellucida by blockage of -SH groups. *Fertil. Steril.* **26,** 201–202.

Reyes, A., Oliphant, G., and Brackett, B. G. (1975). Partial purification and identification of a reversible decapacitation factor from rabbit seminal plasma. *Fertil. Steril.* **26,** 148–157.

Ringertz, N. R., Gledhill, B. L., and Darzynkiewicz, Z. (1970). Changes in deoxyribonucleoprotein during spermiogenesis in the bull. *Exp. Cell Res.* **62,** 204–218.

Rogers, B. J., and Morton, B. (1973a). ATP levels in hamster spermatozoa during capacitation *in vitro*. *Biol. Reprod.* **9,** 361–369.

Rogers, B. J., and Morton, B. E. (1973b). The release of hyaluronidase from capacitating hamster spermatozoa. *J. Reprod. Fertil.* **35,** 477–487.

Rogers, J., and Yanagimachi, R. (1975a). Release of hyaluronidase from guinea pig spermatozoa through an acrosome reaction initiated by calcium. *J. Reprod. Fertil.* **44,** 135–138.

Rogers, J., and Yanagimachi, R. (1975b). Retardation of guinea pig sperm acrosome reaction by glucose: The possible importance of pyruvate and lactate metabolism in capacitation and the acrosome reaction. *Biol. Reprod.* **13,** 568–575.

Roomans, G. M., and Afzelius, B. A. (1975). Acrosome vesiculation in human sperm. *J. Submicr. Cytol.* **7,** 61–69.

Ross, R. N., and Graves, C. N. (1974). O_2 levels in the female genital tract. *J. Anim. Sci.* **39,** 994.

Roussel, J. D., Stallcup, O. T., and Austin, C. R. (1967). Selective phagocytosis of spermatozoa in the epididymis of bulls, rabbits and monkeys. *Fertil. Steril.* **18,** 509–516.

Rubenstein, B. B., Strauss, H., Lazarus, M. L., and Hankin, H. (1951). Sperm survival in women: Motile sperm in the fundus and tubes of surgical cases. *Fertil. Steril.* **2,** 15–19.

Rubin, R. P. (1970). The role of calcium in the release of neurotransmitter substances and hormones. *Pharmacol. Rev.* **22,** 389–418.

Rucklebusch, Y. (1975). Relationship between the electrical activity of the oviduct and the uterus of the rabbit *in vivo*. *J. Reprod. Fert.* **45,** 73–82.

Rumery, R. E., and Eddy, E. M. (1974). Scanning electron microscopy of the fimbriae and ampullae of rabbit oviducts. *Anat. Rec.* **178,** 83–102.

Saksena, S. K., and Harper, M. J. K. (1975). Relationship between concentration of prostaglandin F (PGF) in the oviduct and egg transport in rabbits. *Biol. Reprod.* **13,** 68–76.

Salisbury, G. W., and VanDemark, N. L. (1961). *Physiology of Reproduction and Artificial Insemination of Cattle.* W. H. Freeman, San Francisco.

Sawicki, W., and Koprowski, H. (1971). Fusion of rabbit spermatozoa with somatic cells cultivated *in vitro*. *Exp. Cell Res.* **66,** 145–151.

Schaeffer, G. (1974). A molecular basis for interaction of biguanides with the mitochondrial membrane. In *Biomembranes* (L. Packer ed.), pp. 231–259. Academic Press, New York.

Schaffer, S. W., Safer, B., Scarpa, A., and Williamson, J. R. (1974). Mode of action of the calcium ionophores X-537A and A23187 on cardiac contractility. *Biochem. Pharmacol.* **23,** 1609–1617.

Schill, W. -B., Heimburger, N., Schiessler, H., Stolla, R., and Fritz, H. (1975). Reversible attachment and localization of the acid-stable seminal plasma acrosin–trypsin inhibitors on boar spermatozoa as revealed by the indirect immunofluorescent staining technique. *Z. Physiol. Chem.* **356,** 1473–1476.

Schleuning, W. -D., Hell, R., and Fritz, H. (1976). Multiple forms of boar acrosin and their relationship to proenzyme activation. *Z. Physiol. Chem.* **357,** 207–212.

Schuel, H., Kelly, J. W., Berger, E. R., and Wilson, W. L. (1974). Sulfated acid mucopolysaccharides in the cortical granules of eggs. *Exp. Cell Res.* **88,** 24–30.

Schuel, H., Wilson, W. L., Chen, K., and Lorand, L. (1973). A trypsin-like proteinase localized in cortical granules isolated from unfertilized sea urchin eggs by zonal centrifugation. Role of the enzyme in fertilization. *Dev. Biol.* **34,** 175–186.

Schuetz, A. (1975). Induction of nuclear breakdown and meiosis in *Spisula solidissima* oocytes by calcium ionophore. *J. Exp. Zool.* **191,** 433–440.

Scott, T. W., Voglmayr, J. K., and Setchell, B. P. (1967). Lipid composition and metabolism in testicular and ejaculated ram spermatozoa. *Biochem. J.* **102,** 456–461.

Seiguer, A. C., and Castro, A. E. (1972). Electron microscopic demonstration of arylsulfatase activity during acrosome formation in the rat. *Biol. Reprod.* **7,** 31–42.

Seitz, H. M., Rocha, G., Brackett, B. G., and Mastroianni, L. (1970). Influence of the oviduct on sperm capacitation in the rabbit. *Fertil. Steril.* **21,** 325–328.

Sellens, M. H., and Jenkinson, E. J. (1975). Permeability of the mouse zona pellucida to immunoglobulin. *J. Reprod. Fertil.* **42,** 153–157.

Seshachar, B. R., and Bagga, S. (1963). Cytochemistry of the oocyte of *Loris tardigradus lydekkerianus* and *Macaca mulatta. J. Morphol.* **113,** 119–129.

Settlage, D. S. F., Motoshima, M., and Tredway, D. R. (1973). Sperm transport from the external cervical os to the fallopian tubes in women: A time and quantitation study. *Fertil. Steril.* **24,** 655–661.

Shahani, S. K., Padbidri, J. R., and Rao, S. A. (1972). Immunobiological studies with the reproductive organs, adrenals and spleen of the female mouse. *Int. J. Fertil.* **17,** 161–165.

Sharon, N., and Lis, H. (1972). Lectins: Cell-agglutinating and sugar-specific proteins. *Science* **177,** 949–959.

Shenk, S. L. (1878). Das Saugetierei Künstlich befruchtet ausserhalb des Muttertieves. *Mitt Embryol. Inst. Univ. Wien,* 1877–1885.

Shivers, C. A. (1974). Immunological interference with fertilization. In *Immunological Approaches to Fertility Control* (E. Diczfalusy, ed.), pp. 223–244. Karolinska Institutet.

Shivers, C. A., and Dudkiewicz, A. B. (1974). Inhibition of fertilization with specific antibodies. In *Physiology and Genetics of Reproduction* (E. M. Coutino and F. Fuchs, eds.) part B, pp. 81–96. Plenum Press, New York.

Shivers, C. A., Dudkiewicz, A. B., Franklin, L. E., and Fussel, E. N. (1972). Inhibition of sperm–egg interaction by specific antibody. *Science* **178,** 1211–1213.

Singer, C. (1950). *A History of Biology*. Schuman, New York.

Singer, S. J. (1974). The molecular organization of membranes. *Ann. Rev. Biochem.* **43,** 805–833.

Singer, S. J., and Nicolson, G. L. (1972). The fluid mosaic model of the structure of cell membranes. *Science* **175,** 720–731.

Smith, L. D., and Ecker, R. E. (1971). The interaction of steroids with *Rana pipiens* oocytes in the induction of maturation. *Dev. Biol.* **25,** 232–247.

Smithberg, M. (1953). The effect of different proteolytic enzymes on the zona pellucida of mouse ova. *Anat. Rec.* **117,** 554.

Sobotta, J. (1895). Die Befruchtung und Furchung des Eies der Maus. *Arch. Mikr. Anat.* **45,** 15–93.

Sobotta, J., and Burckhard, G. (1910). Reifung and Befruchtung des Eies der weissen Ratte. *Anat. Hefte,* **42,** 433–497.

Soderwall, A. L., and Blandau, R. J. (1941). The duration of the fertilizing capacity of spermatozoa in genital tract of the rat. *J. Exp. Zool.* **88,** 55–63.

Soderwall, A. L., and Young, W. C. (1940). The effect of aging in the female genital tract on the fertilizing capacity of guinea pig spermatozoa. *Anat. Rec.* **78,** 19–29.

Solter, D., Biczysko, W., Graham, C. F., Pienkowski, M., and Koprowski, H. (1974). Ultrastructure of early development of mouse parthenogenones. *J. Exp. Zool.* **188,** 1–24.

Soupart, P., and Clewe, T. H. (1965). Sperm penetration of rabbit zona pellucida inhibited by treatment of ova with neuraminidase. *Fertil. Steril.* **16,** 677–689.

Soupart, P., and Morgenstern, L. L. (1973). Sperm capacitation and *in vitro* fertilization. *Fertil. Steril.* **24,** 462–478.

Soupart, P., and Noyes, R. W. (1964). Sialic acid as a component of the zona pellucida of the mammalian ovum. *J. Reprod. Fertil.* **8,** 251–253.

Spindle, A. I., and Goldstein, L. S. (1975). Induced ovulation in mature mice and developmental capacity of the embryos *in vitro*. *J. Reprod. Fertil.* **44,** 113–116.

Srivastava, P. N., Adams, C. E., and Hartree, E. F. (1965). Enzymic action of acrosomal preparation on the rabbit ovum *in vitro*. *J. Reprod. Fertil.* **10,** 61–67.

Srivastava, P. N., Zaneveld, L. J. D., and Williams, W. L. (1970). Mammalian sperm neuraminidases. *Biochem. Biophys. Res. Commun.* **39,** 575–582.

Stambaugh, R., and Buckley, J. (1969). Identification and subcellular localization of the enzymes effecting penetration of the zona pellucida by rabbit spermatozoa. *J. Reprod. Fertil.* **19,** 423–432.

Stambaugh, R., and Buckley, J. (1972). Histochemical subcellular localization of the acrosomal proteinase effecting dissolution of the zona pellucida using fluorescein-labeled inhibitors. *Fertil. Steril.* **23,** 348–352.

Stambaugh, R., and Smith, M. (1974). Amino acid content of rabbit acrosomal proteinase and its similarity to human trypsin. *Science,* **186,** 745–746.

Stambaugh, R., Brackett, B. G., and Mastroianni, L. (1969). Inhibition of *in vitro* fertilization of rabbit ova by trypsin inhibitors. *Biol. Reprod.* **1,** 223–227.

Stambaugh, R., Seitz, H. M., and Mastroianni, L. (1974). Acrosomal proteinase inhibitors in rhesus monkey (*Macaca mulatta*) oviduct fluid. *Fertil. Steril.* **25,** 352–357.

Stambaugh, R., Smith, M., and Foltas, S. (1975). An organized distribution of acrosomal proteinase in rabbit sperm acrosomes. *J. Exp. Zool.* **193,** 119–112.

Starke, N. C. (1949). The sperm picture of rams of different breeds as an indication of their fertility. II. The rate of sperm travel in the genital tract of the ewe. *Onderstepoort J. Vet. Sci. Anim. Husb.* **22,** 415–525.

Stegner, H. E., and Wartenberg, H. (1961). Elektronenmikroskopische und histotopochemische untersuchungen über struktur und bildung der zona pellucida menschlicher eizellen. *Z. Zellforsch.* **53,** 702–713.

Steinhardt, R. A., and Epel, D. (1974). Activation of sea urchin eggs by a calcium ionophore. *Proc. Nat. Acad. Sci.* **71,** 1915–1919.

Steinhardt, R. A., Lundin, L., and Mazia, D. (1971). Bioelectric responses of the echinoderm egg to fertilization. *Proc. Nat. Acad. Sci.* **68,** 2426–2430.

Steinhardt, R. A., Epel, D., Carroll, E. J., and Yanagimachi, R. (1974). Is calcium ionophore a universal activator for unfertilized eggs? *Nature (London)* **252,** 41–43.

Stevens, L. C., and Varnum, D. S. (1974). The development of teratomas from parthenogenetically activated ovarian mouse eggs. *Dev. Biol.* **37,** 369–380.

Summers, R. G., Hylander, B. L., Colwin, L. H., and Colwin, A. I. (1975). The functional anatomy of the echinoderm spermatozoon and its interaction with the egg at fertilization. *Amer. Zool.* **15,** 523–551.

Sumner, A. T., Robinson, J. A., and Evans, H. J. (1971). Distinguishing between X, Y and YY-bearing human spermatozoa by fluorescence and DNA content. *Nature New Biol.* **229,** 231–233.

Susi, F. R., and Clermont, Y. (1970). Fine structural modifications of the rat chromatoid body during spermiogenesis. *Amer. J. Anat.* **129,** 177–191.

Szollosi, D. (1966). Time and duration of DNA synthesis in rabbit eggs after sperm penetration. *Anat. Rec.* **154,** 209–212.

Szollosi, D. (1967). Development of cortical granules and the cortical reaction in rat and hamster eggs. *Anat. Rec.* **159,** 431–446.

Szollosi, D. (1972). Changes of some cell organelles during oogenesis in mammals. In *Oogenesis* (J. D. Biggers and A. W. Schuetz, eds.), pp. 47–64. University Park Press, Baltimore.

Szollosi, D. (1975). Mammalian eggs aging in the fallopian tubes. In *Aging Gametes* (R. J. Blandau, ed.), pp. 98–121. Karger, Basel.

Szollosi, D., and Hunter, R. H. F. (1973). Ultrastructural aspects of fertilization in the domestic pig: Sperm penetration and pronucleus formation. *J. Anat.* **116,** 181–206.

Tachi, S., and Kraicer, P. F. (1967). Studies on the mechanism of nidation. XXVII. Sperm-derived inclusions in the rat blastocyst. *J. Reprod. Fertil.* **14,** 401–405.

Talbot, P., Franklin, L. E., and Fussell, E. N. (1974). The effect of the concentration of golden hamster spermatozoa on the acrosome reaction and egg penetration *in vitro*. *J. Reprod. Fertil.* **36,** 429–432.

Tarkowski, A. K. (1971). Recent studies on parthenogenesis in the mouse. *J. Reprod. Fertil., Suppl.* **14,** 31–39.

Tarkowski, A. K., Witkowska, A., and Nowicka, J. (1970). Experimental parthenogenesis in the mouse. *Nature* (*London*) **226,** 162—165.

Terner, C., Maclaughlin, J., and Smith, B. R. (1975). Changes in lipase and phospholipase activities of rat spermatozoa in transit from the caput to the cauda epididymis. *J. Reprod. Fertil.* **45,** 1–8.

Thibault, C. (1947). La parthénogènese experimentale chez le lapin. *C. R. Acad. Sci. Ser. D:* pp. 297–299.

Thibault, C. (1969). *In vitro* fertilization of the mammalian egg. In *Fertilization* (C. Metz and A. Monroy, eds.), Vol. 2, pp. 405–435. Academic Press, New York.

Thibault, C. (1973). *In vitro* maturation and fertilization of rabbit and cattle oocytes. In *The Regulation of Mammalian Reproduction* (S. J. Segal, R. Crozier, P. A. Corfman, and P. G. Condliffe, eds.), pp. 231–240. Charles C Thomas, Springfield, Ill.

Thibault, C., Gerard, M., and Menezo, Y. (1975). Acquisition par l'ovocyte de lapine et de veau du facteur de décondensation du noyau der sparmatozoïde fécondant (MPGF). *Ann. Biol. Anim. Biochem. Biophys.* **15,** 705–714.

Thompson, R. S., and Zamboni, L. (1974). Phagocytosis of supernumery spermatozoa by two-cell mouse embryos. *Anat. Rec.* **178,** 3–14.

Thompson, R. S., Moore-South, D., and Zamboni, L. (1974). Fertilization of mouse ova *in vitro*: An electron microscopic study. *Fertil. Steril.* **25,** 222–249.

Thorn, N. A., and Petersen, O. H., eds. (1975). *Secretory Mechanisms of Exocrine Glands.* Academic Press, New York.

Toyoda, Y., Chang, M. C. (1974). Fertilization of rat eggs *in vitro* by epididymal spermatozoa and the development of eggs following transfer. *J. Reprod. Fertil.* **36,** 9–22.

Toyoda, Y., Yokoyama, M., and Hosi, T. (1971). Studies on the fertilization of mouse eggs *in vitro*. *Jap. J. Reprod.* **16,** 147–151, 152–157.

Troll, W., Schuel, H., and Wilson, W. L. (1974). Induction of polyspermic fertilization of *Arbacia* eggs by specific protease inhibitors leupeptin and antipain. *Biol. Bull.* **147,** 502.

Trounson, A. O., and Moore, N. W. (1974). The survival and development of sheep eggs following complete or partial removal of the zona pellucida. *J. Reprod. Fertil.* **41,** 97–105.

Tsafriri, A., and Channing, C. P. (1975). An inhibitory influence of granulosa cells and follicular fluid upon porcine oocyte meiosis *in vitro*. *Endocrinology* **96,** 922–927.

Tsafriri, A., Lindner, H. R., Zor, U., and Lamprecht, S. A. (1972). *In vitro* induction of meiotic division in follicle-enclosed rat oocytes by LH, cyclic AMP and prostaglandin E_2. *J. Reprod. Fertil.* **31,** 39–50.

Tsafriri, A., Pomerantz, S. H., and Channing, C. P. (1975). Inhibition of oocyte maturation by porcine follicular fluid. *Abstract No. 73, Annual Meeting, Soc. Study Reprod.* Fort Collins, Colo.

Tsafriri, A., Pomerantz, S. H., and Channing, C. P. (1976). Inhibition of oocyte maturation by porcine follicular fluid: Partial characterization of the inhibitor. *Biol. Reprod.* **14,** 511–516.

Tsunoda, Y., and Chang, M. C. (1975a). Penetration of mouse eggs *in vitro*: Optimal sperm concentration and minimal number of spermatozoa. *J. Reprod. Fertil.* **44,** 139–142.

Tsunoda, Y., and Chang, M. C. (1975b). *In vitro* fertilization of rat and mouse eggs by ejaculated sperm and the effect of energy sources on *in vitro* fertilization of rat eggs. *J. Exp. Zool.* **193,** 79–86.

Tsunoda, Y., and Chang, M. C. (1976a). *In vivo* and *in vitro* fertilization of hamster, rat and mouse eggs after treatment with anti-hamster ovary antiserum. *J. Exp. Zool.* **195,** 409–416.

Tsunoda, Y., and Chang, M. C. (1976b). Effect of anti-rat ovary antiserum on the fertilization of rat, mouse and hamster eggs *in vivo* and *in vitro*. *Biol. Reprod.* **14,** 354–361.

Tyrode, M. V. (1910). The mode of action of some purgative salts. *Arch. Int. Pharmacodyn.* **20,** 205–223.

Vacquier, V. D. (1975a). Calcium activation of estero-proteolytic activity obtained from sea urchin egg cortical granules. *Exp. Cell Res.* **90,** 454–456.

Vacquier, V. D. (1975b). The isolation of intact cortical granules from sea urchin eggs: Calcium ions trigger granule discharge. *Dev. Biol.* **43,** 62–74.

Vacquier, V. D., Epel, D., and Douglas, L. A. (1972a). Sea urchin eggs release protease activity at fertilization. *Nature (London)* **237,** 34–36.

Vacquier, V. D., Tegner, M. J., and Epel, D. (1972b). Protease activity establishes the block against polyspermy in sea urchin eggs. *Nature (London)* **240,** 352–353.

Vacquier, V. D., Tegner, M. J., and Epel, D. (1973). Protease released from sea urchin eggs at fertilization alters the vitelline layer and aids in preventing polyspermy. *Exp. Cell Res.* **80,** 111–119.

Vacquier, V. D., and O'Dell, D. S. (1975). Concanavalin A inhibits the dispersion of the cortical granule contents of sand dollar eggs. *Exp. Cell Res.* **90,** 465–471.

Vaidya, R. A., Glass, R. H., Dandekar, P., and Johnson, K. (1971). Decrease in the electrophoretic mobility of rabbit spermatozoa following intrauterine incubation. *J. Reprod. Fertil.* **24,** 299–301.

van Benedin, E. (1875). La maturation de l'oeuf, la fécondation et les premières phases du developpement embryonnaire des mammifères d'apres des recherches faites le lapin. *Bull. Acad. Belg. Cl. Su.* **40,** 686.

VanDemark, N. L., and Hays, R. L. (1954). Rapid sperm transport in the cow. *Fertil. Steril.* **5,** 131–137.

Vandeplassche, M., and Paredis, F. (1948). Preservation of the fertilizing capacity of bull semen in the genital tract of the cow. *Nature (London)* **162,** 813.

Virutamasen, P., Smitasiri, Y., and Fuchs, A.-R. (1976). Intra-ovarian pressure changes during ovulation in rabbits. *Fertil. Steril.* **27,** 188–196.

von Baer, K. (1827). *De Ovi Mammalium et Hominis Genesi.* Leipzig.

von Bregulla, K., Gerlach, U., and Hahn, R. (1974). Versuche zur extrakorporalen reifung, befruchtung und embryonenzucht mit rinderkeimzellen. *Dtsch. Tierarztl. Wochenschr.* **81,** 465–470.

Von der Borch, S. M. (1967). Abnormal fertilization of rat eggs after injection of substances in the ampullae of the fallopian tubes. *J. Reprod. Fertil.* **14,** 465–468.

Wagner, T. E., Mann, D. R., and Vincent, R. C. (1974). The role of disulfide reduction in chromatin release from equine sperm. *J. Exp. Zool.* **189,** 387–393.

Warren, M. R. (1938). Observations on the uterine fluid of the rat. *Amer. J. Physiol.* **122,** 602–608.

Wasserman, N. J., and Masui, Y. (1976). A cytoplasmic factor promoting oocyte maturation: Its extraction and preliminary characterization. *Science* **191,** 1266–1268.

Weil, A. J., and Redenburg, J. M. (1962). The seminal vesicle as the source of the spermatozoa-coating antigen of seminal plasma. *Proc. Soc. Exp. Biol. Med.* **109,** 567–570.

Weinman, D. E., and Williams, W. L. (1964). Mechanism of capacitation of rabbit spermatozoa. *Nature (London)* **203,** 423–424.

Weiss, L. (1969). The cell periphery. *Int. Rev. Cytol.* **26,** 63–105.

Wendt, V., Leidl, W., and Fritz, H. (1975). The lysis effect of bull spermatozoa on gelatin substrate film: Methodological investigations. *Z. Physiol. Chem.* **356,** 315–323.

White, I. G. (1973). Metabolism of spermatozoa with particular relation to the epididymis. *Adv. Biosci.* **10,** 157–168.

Whiteley, A. H., and Chambers, E. L. (1966). Phosphate transport in fertilized sea urchin eggs. II. Effects of metabolic inhibitors and studies on differentiation. *J. Cell Physiol.* **68,** 309–324.

Whittingham, D. G. (1968). Fertilization of mouse eggs *in vitro*. *Nature (London)* **220,** 592–593.

Williams, J. A., and Chandler, D. (1975). Ca^{++} and pancreatic amylase release. *Fed. Proc.* **228,** 1729–1732.

Williams, W. L., Abney, T. O., Chernoff, H. H., Dukelow, W. R., and Pinsker, M. C. (1967). Biochemistry and physiology of decapacitation factor. *J. Reprod. Fertil. Suppl.* **2,** 11–21.

Wimsatt, W. A. (1944). Further studies on the survival of spermatozoa in the female reproductive tract of the bat. *Anat. Rec.* **88,** 193–204.

Witkowska, A. (1973a). Parthenogenetic development of mouse embryos *in vivo*. I. Preimplantation development. *J. Embryol. Exp. Morphol.* **30,** 519–545.

Witkowska, A. (1973b). Parthenogenetic development of mouse embryos *in vivo*. II. Postimplantation development. *J. Embryol. Exp. Morphol.* **30,** 547–560.

Wobschall, D., and Ohki, S. (1973). Electrical polarization of phosphatidylserine bilayer membranes by calcium ions. *Biochim. Biophys. Acta* **291,** 363–370.

Wolff, C. (1759). *Theoria Generationes*. Halle.

Wunderlich, F., Müller, R., and Speth, V. (1973). Direct evidence for a colchicine-induced impairment in the mobility of membrane components. *Science* **182,** 1136–1138.

Wyrick, R. E., Nishihara, T., and Hedrick, J. L. (1974). Agglutination of jelly coat and cortical granule components and the block to polyspermy in the amphibian *Xenopus laevis*. *Proc. Nat. Acad. Sci.* **71,** 2067–2071.

Yamanaka, H. S., and Soderwall, A. L. (1960). Transport of spermatozoa through the female genital tract of hamsters. *Fertil. Steril.* **11,** 470–492.

Yanagimachi, R. (1964). The behavior of hamster sperm to the hamster and mouse ova *in vitro*. *Proc. Vth Int. Congr. Anim. Reprod. Artif. Insem.* **7,** 292.

Yanagimachi, R. (1966). Time and process of sperm penetration into hamster ova *in vivo* and *in vitro*. *J. Reprod. Fertil.* **11,** 359–370.

Yanagimachi, R. (1969a). *In vitro* capacitation of hamster spermatozoa by follicular fluid. *J. Reprod. Fertil.* **18,** 275–286.

Yanagimachi, R. (1969b). *In vitro* acrosome reaction and capacitation of golden hamster spermatozoa by bovine follicular fluid and its fractions. *J. Exp. Zool.* **170,** 269–280.

Yanagimachi, R. (1970a). *In vitro* capacitation of golden hamster spermatozoa by homologous and heterologous blood sera. *Biol. Reprod.* **3,** 147–153.

Yanagimachi, R. (1970b). The movement of golden hamster spermatozoa before and after capacitation. *J. Reprod. Fertil.* **23,** 193–196.

Yanagimachi, R. (1972a). Fertilization of guinea pig eggs *in vitro*. *Anat. Rec.* **174,** 9–20.

Yanagimachi, R. (1972b). Penetration of guinea pig spermatozoa into hamster eggs *in vitro*. *J. Reprod. Fertil.* **28,** 477–480.

Yanagimachi, R. (1975). Acceleration of the acrosome reaction and activation of guinea pig spermatozoa by detergents and other reagents. *Biol. Reprod.* **13,** 519–526.

Yanagimachi, R., and Chang, M. C. (1961). Fertilizable life of golden hamster ova and their morphological changes at the time of losing fertilizability. *J. Exp. Zool.* **148,** 185–197.

Yanagimachi, R., and Chang, M. C. (1963a). Sperm ascent through the oviduct of the hamster and rabbit in relation to the time of ovulation. *J. Reprod. Fertil.* **6,** 413–420.

Yanagimachi, R., and Chang, M. C. (1963b). Infiltration of leucocytes into the uterine lumen of the golden hamster during the oestrus cycle and following mating. *J. Reprod. Fertil.* **5,** 389–396.

Yanagimachi, R., and Chang, M. C. (1964). *In vitro* fertilization of golden hamster ova. *J. Exp. Zool.* **156,** 361–376.

Yanagimachi, R., and Nicolson, G. L. (1974). Changes in lectin-binding to the plasma membrane of hamster eggs during maturation and preimplantation development. *J. Cell Biol.* **63,** 381a.

Yanagimachi, R., and Noda, Y. D. (1970a). Electron microscopic studies of sperm incorporation into the golden hamster egg. *Amer. J. Anat.* **128,** 429–462.

Yanagimachi, R., and Noda, Y. D. (1970b). Ultrastructural changes in the hamster sperm head during fertilization. *J. Ultrastruct. Res.* **31,** 465–485.

Yanagimachi, R., and Noda, Y. D. (1970c). Physiological changes in the postnuclear cap region of the mammalian sperm: A necessary preliminary to the membrane fusion between sperm and egg cells. *J. Ultrastruct. Res.* **31,** 486–493.

Yanagimachi, R., and Noda, Y. D. (1970d). Fine structure of the hamster sperm head. *Amer. J. Anat.* **128,** 367–388.

Yanagimachi, R., and Noda, Y. D. (1972). Scanning electron microscopy of golden hamster spermatozoa before and during fertilization. *Experientia* **28,** 69–72.

Yanagimachi, R., and Usui, N. (1972). The appearance and disappearance of factors involved in sperm chromatin decondensation in the hamster egg. *J. Cell Biol.* **55,** 293a.

Yanagimachi, R., and Usui, N. (1974). Calcium dependence of the acrosome reaction and activation of guinea pig spermatozoa. *Exp. Cell Res.* **89,** 161–174.

Yanagimachi, R., Noda, Y. D., Fujimoto, M., and Nicolson, G. L. (1972). The distribution of negative surface charges on mammalian spermatozoa. *Amer. J. Anat.* **135,** 497–520.

Yanagimachi, R., Nicolson, G. L., Noda, Y. D., and Fujimoto, M. (1973). Electron microscopic observations of the distribution of acidic anionic residues on hamster spermatozoa and eggs before and during fertilization. *J. Ultrastruct. Res.* **43,** 344–353.

Yang, W. H., Lin, L. L., Wang, J. R., and Chang, M. C. (1972). Sperm penetration through zona pellucida and perivitelline space in the hamster. *J. Exp. Zool.* **179,** 191–206.

Yang, C. H., and Srivastava, P. N. (1974a). Purification and properties of aryl sulfatases from rabbit sperm acrosomes. *Proc. Soc. Exp. Biol. Med.* **145,** 721–725.

Yang, C. H., and Srivastava, P. N. (1974b). Separation and properties of hyaluronidase from ram sperm acrosomes. *J. Reprod. Fertil.* **37,** 17–25.

Zamboni, L. (1970). Ultrastructure of mammalian oocytes and ova. *Biol. Reprod., Suppl.* **2,** 44–63.

Zamboni, L. (1971a). *Fine Morphology of Mammalian Fertilization.* Harper & Row, New York.

Zamboni, L. (1971b). Acrosome loss in fertilizing mammalian spermatozoa: A Clarification. *J. Ultrastruct. Res.* **34,** 401–405.

Zamboni, L. (1972a). Fertilization in the mouse. In *Biology of Mammalian Fertilization and Implantation* (K. S. Moghissi and E. S. E. Hafez, eds.), pp. 213–262. Charles C Thomas, Springfield, Ill.

Zamboni, L. (1972b). Comparative studies on the ultrastructure of mammalian oocytes. In *Oogenesis* (J. D. Biggers and A. N. Schuetz, eds.), pp. 5–45. University Park Press, Baltimore.

Zamboni, L., Zemjanis, R., and Stefanini, M. (1971). The fine structure of monkey and human spermatozoa. *Anat. Rec.* **169,** 129–153.

Zamboni, L., Moore-Smith, D., and Thompson, R. S. (1972). Migration of follicle cells through the zona pellucida and their sequestration by human oocytes *in vitro*. *J. Exp. Zool.* **181,** 319–340.

Zaneveld, L. J. D., and Williams, W. L. (1970). A sperm enzyme that disperses the corona radiata and its inhibition by decapacitation factor. *Biol. Reprod.* **2,** 363–368.

Zaneveld, L. J. D., Robertson, R. T., and Williams, W. L. (1970). Synthetic enzyme inhibitors as antifertility agents. *Fed. Eur. Biochem. Soc. Lett.* **11,** 345–347.

Zaneveld, L. J. D., Robertson, R. T., Kessler, M., and Williams, W. L. (1971). Inhibition of fertilization *in vivo* by pancreatic and seminal plasma trypsin inhibitors. *J. Reprod. Fertil.* **25,** 387–392.

Zaneveld, L. J. D., Polakoski, K. L., and Williams, W. L. (1972). Properties of a proteolytic enzyme from rabbit sperm acrosomes. *Biol. Reprod.* **6,** 30–39.

Zelenin, A. V., Shapiro, I. M., Kolesnikov, V. A., and Senin, V. M. (1974). Physico-chemical properties of chromatin of mouse sperm nuclei in heterokaryons with Chinese hamster cells. *Cell Differ.* **3,** 95–101.

Index

Acrosin
  inactivation of, 63
  inhibitors of, 22-24, 83
  location in sperm, 83
  molecular weight of, 22
  role in zona penetration, 81
Acrosome cap, 63
Acrosome reaction
  conditions for, 65-66
  continuity of membranes following, 65
  induction by ionophores, 67
  induction by serum components, 66
  role of acrosin in, 66
  role of Ca^{2+} in, 66-67
  vesiculation in, 63, 65
Antiovary serum
  effect on implantation of, 79
  inhibition of fertilization by, 79
  specificity of, 78

Binding between gametes, 69-73
Boromycin, 91

Calcium, 40
Capacitation of sperm
  alteration of sperm membrane by, 54-55, 59
  altered motility following, 60
  by blood serum and follicular fluid, 59-60
  by cumulus oophorus, 55-59
  and decapacitation factors, 60
  by β-glucuronidase, 59
  inhibition of, 56, 60
  removal of seminal plasma by, 53
  site of, 55
  time required for, 54
Cat, 38
Centrioles, 5
Chinese hamster, 37
Cortical reaction
  induction of, 91-93
  membrane depolarization in, 92
  reorganization of plasma membrane in, 93
  role of Ca^{2+} in, 93
  role of sperm in, 91
Cow, 38
Cross-linking, 99-101
Cumulus oophorus, 10-12

Dog, 39

Egg
  aging of, 12
  cortical granules of, 8
  dictyate stage, 5
  life span after ovulation, 28-29
  maturation of, 8-10
  microvilli of, 87-88
  mitochondria of, 8
  origin of, 5
  ovulation of, 27
  transport of, 27-29
  uptake of macromolecules by, 6
Electrical stimulation of eggs, 91-92, 115
Epididymis, 14

Fertilization cone, 84, 88
Follicle cells, 6
Fusion of gametes
  egg maturation and, 89
  fertilization cone and, 84, 88
  intermixing of membrane components in, 88
  non-species specificity of, 89
  role of capacitation in, 89
  role of egg microvilli in, 87
  role of postacrosomal membrane in, 87

Glycosidases, 56-59, 99
Golden hamster, 36-37, 42-51
Golgi complex, 5
Guanidines, 91
Guinea pig, 37

History of fertilization, 2-3
Hyaluronidase, 19, 61
Hydrogen-ion concentration, 40

In vitro fertilization, 34-51
Ionophores, 91, 93-95, 115

Lectins, 78, 92-93, 99

Man, 38
Metabolic activation, 109-112
Mouse, 35-36
Neuraminidase, 91, 93

Oogenesis, 5
Oogonia, 5
Oviduct
  isthmus of, 30, 32
  utero-tubal junction of, 30

Parthenogenesis
  Ca^{2+} and, 118
  cortical reaction and, 118-119
  cumulus removal and, 115-118
  death after, 119
  electrical stimulation and, 115-116
  ionophores and, 115, 118
  membrane changes in, 118
  as model for malignancy, 124
  osmotic shock and, 115
  temperature shock and, 115
  teratomas and, 119
  types of, 116
Penetration of eggs by sperm
  angle of, 83
  egg maturation and, 85
  pH and, 85
  role of acrosin in, 81-83
  time required for, 85
Polyspermy prevention
  cortical reaction in, 91-95
  egg aging and, 101
  medium and, 101-102
  vitelline reaction in, 95-96
  zona reaction in, 96-102
Practical applications, 123-124
Proacrosin, 21
Pronucleus formation
  chromatin in, 103-105

Pronucleus formation (*cont'd*)
- DNA synthesis in, 105
- egg maturation and, 105-107
- male pronucleus growth factor in, 107
- non-species specificity of, 105
- nuclear membrane in, 103-104
- nucleoli in, 103

Rabbit, 34
Rat, 35
Receptor on sperm, 79
Receptor on zona
- effect of enzymes on, 75
- heat stability of, 76
- inhibition of fertilization by, 76-77
- on inside of zona, 76
- masking of lectins, 78
- solubilization of, 76-77

Seminal plasma
- coagulation of, 30
- coating antigens of, 25

Somatic cells, 113-114
Sperm receptor hydrolase, 98-99
Spermatozoa
- acrosomes of, 13, 19-21
- acrosome reaction of, 61-67
- antigens of, 15
- composition of, 14
- contractile system of, 24
- cytoplasmic droplet of, 14
- dense fibers of, 24
- DNA of, 13
- enzymes of, 19-22
- F-body of, 14
- interactions with somatic cells, 113-114
- lectin binding by, 15
- microtubules of, 24
- mitochondria of, 24
- nuclear condensation of, 14

Spermatozoa (*cont'd*)
- nutrients for, 24
- perforatorium of, 24, 83-85
- regional differences in plasma membrane, 15-16
- respiration of, 25
- surface of, 15
- surface charge on, 14, 18
- survival in female tract, 31-32
- terminal sugars of, 15
- transport to site of fertilization, 30-32
- wastage of, 32

Squirrel monkey, 39

Transport of gametes, 27-32
Trypsin inhibitors, 81-82

Unsolved problems, 121-123

Vitelline delaminase, 98-99
Vitelline reaction
- lectin binding after, 96
- membrane reorganization in, 95-96
- time required for, 95-96

Zona pellucida
- antigenicity of, 6, 78-79
- binding of lectins by, 6
- composition of, 6
- formation of, 6-8
- penetration by sperm, 81-85
- permeability of, 6-8
- receptor for sperm in, 75-78
- thickness of, 6

Zona reaction
- block by trypsin inhibitors, 97
- cortical granule contents in, 96
- effect of egg aging on, 101
- induction of, 96
- partial hydrolysis of zona in, 98
- production by trypsin, 97-98